PLATIES

KEEPING AND BREEDING THEM IN CAPTIVITY

PHOTO BY ANDRE ROTH.

A sunburst platy variatus male with a golden variatus female. These variatus and other platies have many different trade names. This variety has been called *sunset* platy variatus for thirty years before they changed the name to Sunburst!

by
Donald Mix

INTRODUCTION

For those who are contemplating the keeping of tropical aquarium fishes, the platies are excellent choices for a number of reasons. Not only are they quite hardy, but they readily breed, a big plus for the novice. They are available in a number of hybrid forms and color varieties. There is certainly no shortage of types from which to select or to specialize in.

Along with the guppies and mollies, the platies comprise a large part of the aquatic hobby where tropical species are concerned. Because they are so popular and therefore common in pet shops, you will not have to pay a high price for your initial stock, though obviously the rarer colors and patterns will be somewhat more expensive in comparison to the readily available forms.

All platies make good community tank inhabitants, so whether you want a mixed collection of fishes, a specialized collection, or just a few fishes to decorate an aquarium essentially devoted to plants, these fishes will meet your needs.

Although this group of species is hardy and easy to maintain, this does not mean they can be abused with regard to the conditions they need to survive. Hardy simply means that they are more flexible in their ability to

PHOTO BY ANDRE ROTH.

These gold platies are also called *moons* because of the black spot on the caudal peduncle, or even *Mickey Mouse* because there are resemblances to this animated character, too.

PHOTO BY ANDRE ROTH.

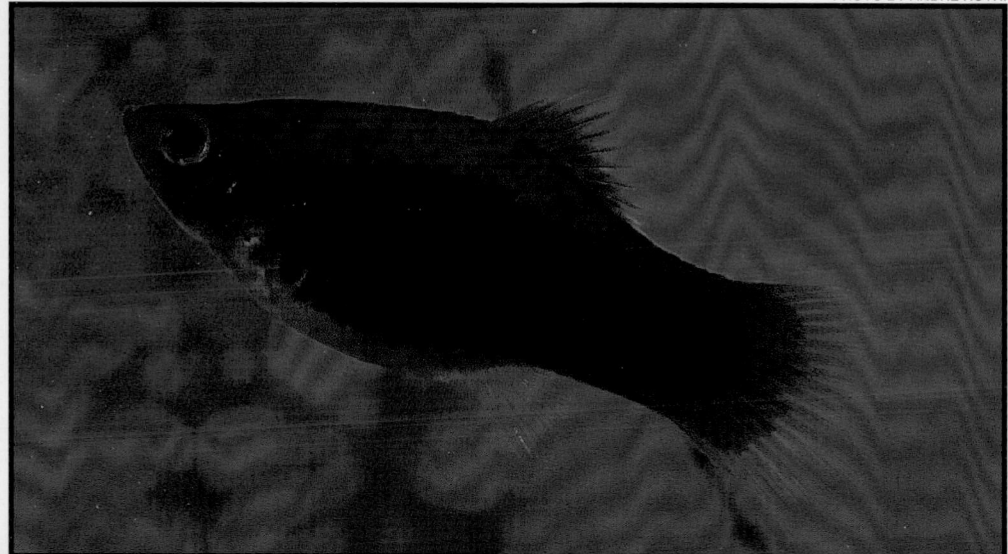

Red tuxedo platy female. The white dot protruding from her vent indicates she is imminently due to drop her young. She has from 25-60 babies (fry) every month.

adjust to a variety of water conditions. But, once established, those conditions must be reasonably constant if problems are to be avoided. Likewise, the very fact that these fishes do procreate readily does present its own set of problems. Unless you are careful you could very soon have far more fishes than you know what to do with. At the same time there are other problems related to their breeding habits that make specialized breeding extremely difficult. More will be said about that in the breeding chapter.

In the following chapters no assumptions are made about your aquatic knowledge. All aspects of fishkeeping are discussed from a basic need-to-know platform. This will mean that you can commence with confidence and with a single reference to guide you. Once underway you can then decide whether or not you wish to delve into this or that area of the hobby in greater depth. You are then referred to the larger and more detailed works published by TFH that will add to the knowledge you have gained from this book.

What should be stressed, and this will be repeated in later chapters, is that you should not rush into any hobby, especially fishkeeping, that has a number of technical requirements associated with it. Exercise control over your desire to get under way as quickly as possible. Careful thought should go into the selection of the tank and all its ancillary equipment. If you proceed cautiously, and with patience, you will avoid many of the problems that first time aquarists often encounter simply because they want to set up their display tank and place fishes into it on the same day.

WHAT ARE PLATIES?

From a hobbyist's standpoint there are two ways in which specific groups of animals can be viewed. You can study them and their scientific standing among all animals, or you can view them from a purely practical viewpoint. By this is meant how they can be accommodated, what they feed on, how they are bred, which types will live with other types, and similar husbandry considerations.

Most hobbyists take this practical route but, as a result, their knowledge of the pets they keep is definitely less than it could be. In the next chapter we will view platies from a practical viewpoint. Here we will consider their basic scientific standing, and aspects related to this. Your knowledge of these fishes will then be broad based, rather than narrow based.

CLASSIFICATION OF ANIMALS

With over one million animals on our planet there was clearly a need to develop a system that divided them into mutually similar groups that could be readily referred to. One way of doing this is by using obvious features that each possesses. For example, fishes live in water, humans and dogs do not. Birds, insects, and bats fly, whereas cats and cattle do not. Such systems can be useful, but create problems when we start to look at the animals more closely. Whales and dolphins look like fishes and live in water, but they are not fishes. Many insects and bats have wings, but they are not birds. Snakes crawl around like worms, but they are not related to them. Obvious features can therefore be misleading. Zoologists developed a system of identification that was based on supposed evolutionary affinities between animals. In taking this approach all features are considered and compared.

For example, a whale breathes air, it possesses hair, its offspring develop with a placenta, and their babies suckle milk from their mother after they are born. When these and many other features are considered it becomes clear that a whale has more in common with a human or a cat and dog, than it does with fishes and other water inhabiting creatures. Likewise, an ostrich cannot fly but, this aside, every other feature clearly indicates it is a bird.

The system of classification used internationally to divide animals into groups is known as the binomial system of nomenclature. It was developed by a Swedish naturalist, Carolus Linnaeus, during the 18th century. In this system all animal organisms are placed into a series of groups, or taxa, based on features they share with other animals. Each group is then divided into further groups using similarities as the basis for the grouping. This is repeated until we eventually arrive at individual species, such as the various wild cats, or bears, or monkeys—or fishes.

whether you are Chinese, Russian, English, or French—a considerable advantage that avoids misunderstanding of what animal is being discussed.

The system of binomial nomenclature is governed by a set of internationally agreed upon rules in order to avoid confusion. The latter is still possible, either by people misusing the system, or simply because its very flexibility to embrace new knowledge means it is a changing system rather than a static one. At our level, we do not need to bother about many of the technicalities of the system, but a few aspects should be

PHOTO BY DR. AXELROD.

This variety of golden platy hybrid was developed by Dr. Herbert R. Axelrod in 1966 but never became popular.

SCIENTIFIC NAMES

The scientific method of classification uses Latin for the group and species names. Today, it also includes Latinized words from other languages. Latin (and Greek) was originally used because it was the language of scholars and was internationally acceptable. Scientific names are thus the same regardless of learned or at least you should be aware of them because you will be confronted with them time and again in books on tropical fishes.

A species is indicated by the use of two names. The first is its genus (generic) name, the second is the species (specific) or trivial name. Only when the generic and trivial names are used together(!) is a species uniquely identified. The generic and specific names always appear in a type face that differs from the main body of text. Thus you will normally see scientific names in *italics*. The generic name always begins with a capital letter, the trivial name always with a lowercase letter. Take the green swordtail as an example. This fish was named by the German Johannes Heckel in 1848 for his fellow countryman Carl Heller. Its generic name is *Xiphophorus* (meaning sword bearer), while the trivial name is *helleri*. The species is thus *Xiphophorus helleri* (the generic name can be abbreviated to its initial letter after it has been previously written in full within a text, thus *X. helleri*).

Subsequent to this, another form of the swordtail was discovered, but was not different enough from the original species to be given its own species name. It can be named as a subspecies. The form on which the species was based then has its trivial name repeated to form a trinomial, thus *X. helleri* becomes *X. helleri helleri*. This

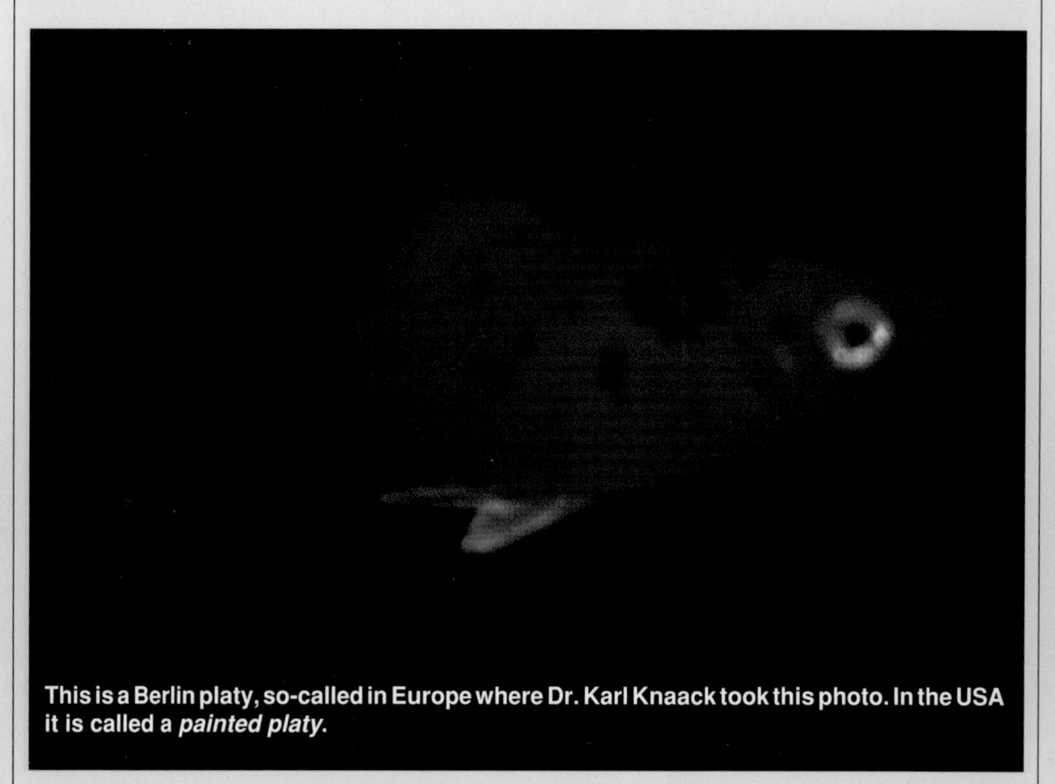

This is a Berlin platy, so-called in Europe where Dr. Karl Knaack took this photo. In the USA it is called a *painted platy*.

tells you it was the original swordtail discovered. The new form is given its own trinomial, thus you would have *X. helleri alvarezi.* Other subspecies may also be discovered and have their own trinomial, such as *X. h. strigatus*, and so on. If the differences between the forms are subsequently considered great enough to consider one or more of them valid species the subspecific name is used as the species name and the "middle" name is dropped. *Xiphophorus helleri alvarezi* would become *X. alvarezi.*

When additional information is discovered about a species it may happen that it is moved into another genus—or it may be placed into another genus by an expert from the outset simply because that expert was not aware that the fish had already been described and named previously, possibly in another country. When this happens it is customary to indicate the former genus thus:

PHOTO BY GENE WOLFSHEIMER.

A pair of golden wagtail platies.

Xiphophorus (= *Mollienesia*) *helleri* in order to avoid confusion. When a fish changes genus it will normally retain its trivial name. The name and year after a scientific name indicates when that name was first applied to the species, and by whom. If the name is in parenthesis this indicates the author placed the fish in a different genus to the one it is now in. For example, the common platy was originally described by Günther in 1866 as *Platypoecilus maculatus* and therefore was written *Platypoecilus maculatus* Günther, 1866. This fish was later transferred to the genus *Xiphophorus* and the name changed to *Xiphophorus maculatus* (Günther, 1866).

ANIMAL GROUPS AND NAMES

Although there are many group names below the all embracing one known as Animalia—the animals as opposed to the plants—certain of these are held to be more important than others. They are regarded as obligatory in the formal classification of any animal. These groups, in descending order, are the phylum, class, order, family, genus, and species.

We can skip the phylum, though for completeness fishes are in the one called Chordata—animals with a central notochord, and the superclass Gnathostomata, the jawed fishes. All bony fishes are placed in the class Osteichthyes, in contrast to the class Chondrichthyes (sharks, rays, and other cartilaginous fishes). In all, there are some 20,000 fish species and, of these, the vast majority are bony fishes. Fishes are the most numerous vertebrate animals, there being 9,000 birds (class Aves), 6,000 reptiles (class Reptilia), 4,500 mammals (class Mammalia), and 2,500 amphibians (class Amphibia).

The major features of a bony fish are that it has a notochord, is jawed, has a bony skeleton, breathes via gills, has paired fins, its body is covered with scales, and it has three semicircular canals.

The class Osteichthys is divided into about 40 orders (very variable depending on the author of the classification). Our interest is in the one called Cyprinodontiformes (the killifishes and their allies). This contains the tooth bearing carps that, generally, are surface feeders. The order is divided into some nine families, one of which is called Poeciliidae—the livebearing tooth carps.

The family Poeciliidae contains the swordtails and platies, as well as their famous relatives the guppy and mollies, plus a number of other species well known to aquarists. The members of this family are relatively recent (44-38 million years old) in their development compared to the egg-laying fishes. They are distributed in the New World, being found in the fresh and brackish (estuarine) waters of the tropical and subtropical Americas and the Caribbean islands.

They are elongate, possess sharp teeth, all fins are supported by soft rays, they have superior (upturned) mouths, and their lateral line (a series of pores connecting to a canal that in turn connects to the inner ear) is evident only in the head region. In most species the sexes are dimorphic, which means that they are readily identified by their external appearance. There are about 20 genera in this family, these housing almost 200 species.

The swordtails and platies are

A beautiful red tuxedo wagtail platy.

grouped together in the genus *Xiphophorus*. At the generic level these fishes are very similar in most respects. Differences are relatively minor and related to aspects of their external form rather than any of a major anatomical or physiological nature. There are about 21 species within this genus. These are conveniently divided into three non-scientific group types by aquarists.

Of all bony fishes only a very few (about 500 or so) are livebearers (termed viviparous). Of these, not all are true livebearers in the sense that the female provides direct nourishment to her offspring. Many are ovoviviparous, which means the eggs are retained within the female and draw nourishment from their egg yolk. They are then released into their watery world in an advanced state, thus giving the appearance of being true livebearers.

The line between the two livebearing types is not clear cut. In many species, including platies, the young receive some food from their mother, but this is not sufficient to totally sustain them. The rest is supplied via their egg yolk. They are thus ovoviviparous rather than true livebearers. Egglaying fishes, by the way, are termed oviparous.

Although swordtails and platies are placed in the same genus today, this was not always the case. Years ago the platies were placed in the genus *Platypoecilus*. However, the discovery of *P. xiphidium* Gordon, 1932, clearly indicated that platies and swordtails were more closely related than was previously thought. Following the discovery of *Xiphophorus milleri* Rosen, 1960, it was decided that all platy species were so closely related to the swordtails that they should be regarded as belonging to the same genus. This being the case, *Xiphophorus* Heckel, 1848, because it was proposed before *Platypoecilus* Günther, 1866, and therefore had "priority," was used to house both the platies and swordtails.

A final comment on this genus. The term swordtail refers indeed to the extension of the lower rays of the caudal fin. The name *Xiphophorus*, or "sword bearer," however, does not refer to the "sword" seen on the tails of the males of many of its members, as might be thought, but refers to the modification of the rays of the anal fin that form the copulatory gonopodium.

COMMON NAMES

The use of common names is not in any way regulated. Any person is free to call a given species whatever they wish. Some common names have been universally accepted through conventional use, others less so. This means that while a species has only one current scientific name it may have any number of common names. Furthermore, two different species may have the same common name, which can clearly create problems. With these facts in mind it is always preferable to be familiar with the scientific names of the species you keep. Hybrids and color variants have no scientific standing (they are not separate species) but invariably have common names.

Rare rainbow platy variatus male which also shows the moon pattern on the caudal peduncle.

PHOTO BY DR. KARL KNAACK.

PHOTO BY DR. KARL KNAACK.

These are red wagtail swordtails. They are closely related to platies. The swordtails have an elongated tail (males only!); they freely interbreed with platies and very few platies in the hobby don't have some swordtail genes.

LIVING IN A WORLD OF WATER

Although the chemical formula H_2O stands for water, it does not indicate what sort of water it is. There are many types, and the aquarist uses these to divide the hobby into its various disciplines.

FRESHWATER AND MARINE FISHKEEPING

The most basic aquatic division is salt water, found in oceans, seas, and even some lakes, and fresh water, found in rivers, lakes, and ponds. Fishes have evolved mechanisms enabling them to live in one or the other. If they are marine they cannot survive in fresh water, and vice versa. This is for the following reason. A marine fish lives in an environment that has a density that is greater than that of its own body fluids. Through a process known as osmosis its body is constantly losing liquids and salts. To overcome this it drinks copiously and excretes little liquid urine. The reverse is true of freshwater species. To overcome the invasion of liquids they drink very little and excrete much urine.

A few fish species are able to survive in both of these water types during the course of their lives—salmon are a classic example. To do this they develop special physiological features at different times of their lives. In other cases, species can survive in brackish waters (neither fully salt- nor freshwater), which are usually found in estuaries. They can survive in pure marine or pure fresh water for short periods of time, but generally favor one or the other. In the case of guppies it is fresh water. Swordtails and platies are full freshwater species.

WATER TEMPERATURE

The water temperature a species has evolved to live in gives rise to a second major aquatic division, tropical or cold water. There is no sharp dividing line between one or the other, because fishes are found at just about every temperature. Each species cannot survive beyond a certain temperature level, others not below a certain level. For practical purposes it can be stated that most "tropical" species require a temperature of about 24°C (75°F), and the tropical range is in the order of 15-30°C (59-86°F).

Certain swordtails and platies can cope with temperatures as low as 16°C (61 °F), but become increasingly unhappy once the temperature goes much below 26°C (78°F). As the temperature increases, so does their appetite, and vice versa. It is always prudent to maintain the temperature within a degree or two of their ideal "comfort" zone, reducing this periodically a degree or two for very short periods. This can have a stimulating effect on fishes,

PHOTO BY AQUAPRESS, MP & C PIEDNOIR.

A tankful of *Xiphophorus maculatus*, red platy hybrids in a perfect tank setting. The fry hide in the bushy plants.

since it mimics the situation that is found seasonally in tropical waters.

ACIDITY AND HARDNESS (pH & DH)

Having established that platies are tropical freshwater fishes, we can now consider other aspects of the water quality they will require. Two very important considerations are the levels of acidity and hardness. If these are not within the required range of the fishes, the effect will be poor health. This is because the various species have become adapted to given amounts of ions and minerals in the water.

If these exceed, or do not meet, their tolerance levels, cell physiology is effected. Metabolic process are slowed down, or accelerated, to the degree that they have a deleterious effect on the fishes in various ways. Breathing may be impaired, excessive skin mucus may be produced, breeding difficulties ensue, and resistance to diseases will be reduced.

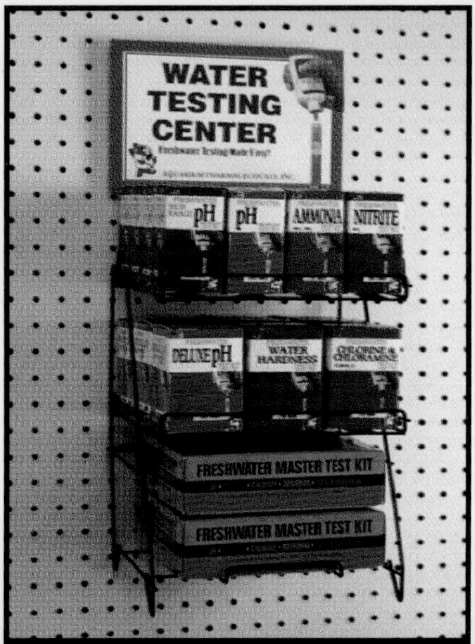

pH: The acidity or alkalinity of water is a measure of the amount of hydrogen ions it contains. The more hydrogen ions there are the more acidic the water is, the less there are the more alkaline it is. Acidity and alkalinity are measured by a pH scale which ranges from 0-14, with 7 being regarded as the neutral point.

The lower the number below 7 the more acidic the water, and the higher the number beyond 7 the more alkaline. A difference of one full number represents a ten-fold change in the relative acidity or alkalinity of the water. Platies are relatively tolerant to the pH as long as it is slightly on the alkaline side of neutral. Their tolerance range is pH 7-8.4 with 7.2-7.5 being about right. Fortunately, this corresponds to that of most domestic water supplies. The pH value is effected by the water temperature, its mineral content, other compounds, and gases such as carbon dioxide, so it does fluctuate during any 24-hour cycle. With this in mind it is wise to take periodic readings at the same time of day.

Your local pet shop should have a water testing center in which various chemical, paper strips and indicator dyes are available with instructions how to use them. Photo courtesy of Aquarium Pharmaceuticals.

Testing for pH is a simple matter using one of the many excellent kits available from your pet or aquatic dealer which come with complete instructions for their use. In most tests, a reagent is added to a sample of the water to be tested which will cause it to change color. A color chart is supplied; comparison of the test water color with this chart indicates the pH. To increase the pH in an aquarium (make it more alkaline) the use of limestone rocks will help, as will the addition of chalk chips to a filter system.

Water Hardness: This may range from very soft to very hard. Generally, acidic water is soft while alkaline water is hard, but this is not always the case. Hardness is a measure of the amount of various elements and compounds in the water. It can be expressed in various ways, but the two most popular methods are German degrees (DH), and parts per million (ppm). Platies prefer water that is slightly to moderately hard, and which will have a DH reading of 12-20, which equates to 216-360 ppm. In order that you can convert any of the systems popular in your country into ppm, and to each other, the following approximations will be useful:

-English° (Clark's) = 14 ppm
-American° = 17 ppm
-German° = 18 ppm

As an example, to convert 12° English to German you would multiply by 14 then divide by 18 to give you 9.3 DH.

In order to determine the hardness in your aquarium you can again use any of the many test kits available from pet and aquatic stores. To increase the hardness use the same method as for increasing alkalinity. To reduce alkalinity or hardness you can pass the water through an organic filter (such as peat) or, more easily, dilute the water by the addition of distilled water (which has a neutral pH value and no hardness). While platies are quite tolerant toward a range of pH and DH conditions, what you must always bear in mind is that this only holds true if the changes are gradual. Sudden changes in water chemistry could be fatal because they give the fishes no time to adjust. The same is equally true of rapid temperature changes much beyond one or two degrees.

CHLORINE

As you are aware, domestic water supplies are variably treated with chlorine and chloramines in order to make the water safe for human consumption. While our bodies can cope with them they are lethal to fishes, which are infinitely smaller than us. Chlorine is easily dissipated into the air by vigorous aeration of the water or by the addition of chlorine eliminating tablets from your pet shop. Chloramines are more stable and must be removed by chemical means—again via tablets from your pet shop. You can purchase test kits for these compounds from your pet dealer.

It should be added with respect to all domestic water supplies that these are never constant. In other words, it is not enough for you to establish the pH, hardness, and chlorine levels of your local supply and assume these values will remain the same. They are changed, for example after flooding, in order to reduce the risk of pathogens in the water, or for other comparable reasons. Routine water quality testing is therefore obligatory for any aquarium enthusiast if problems are to be avoided.

Test kits for copper and other elements are also available, and are recommended if you know you have copper (or lead or iron) pipes dispensing your domestic water. As a precaution, it is always wise to run the tap water for a while before using it to effect partial water changes in your aquarium.

THE NITROGEN CYCLE

When you commence setting up an aquarium it must never be forgotten that you are creating a complete world for its inhabitants. It is a mini environmental ecosystem in which you are trying to match the conditions the fish would live under in their natural habitat.

In such a habitat the full forces of nature are acting upon the water in order to make it a suitable place to live in. One of these forces is the nitrogen cycle, which results in the organic decomposition of dead materials, such as plants, microorganisms, and even large organisms, including the fishes. That which cannot be "processed" via natural biological means is removed by the ever onward motion of a stream or river.

Byproducts of the metabolism of fishes are urine and fecal matter—which are organic and are broken down by decomposition into lethal compounds. This natural biological process changes these lethal organic compounds into less lethal substances. Some of these are absorbed by plants as food. In turn, the plants are eaten by the fishes.

A group of sunburst platies showing the variation within a single litter.

PHOTO BY DR. WARREN E. BURGESS.

PHOTO BY AQUAPRESS, MP & C. PIEDNOIR.

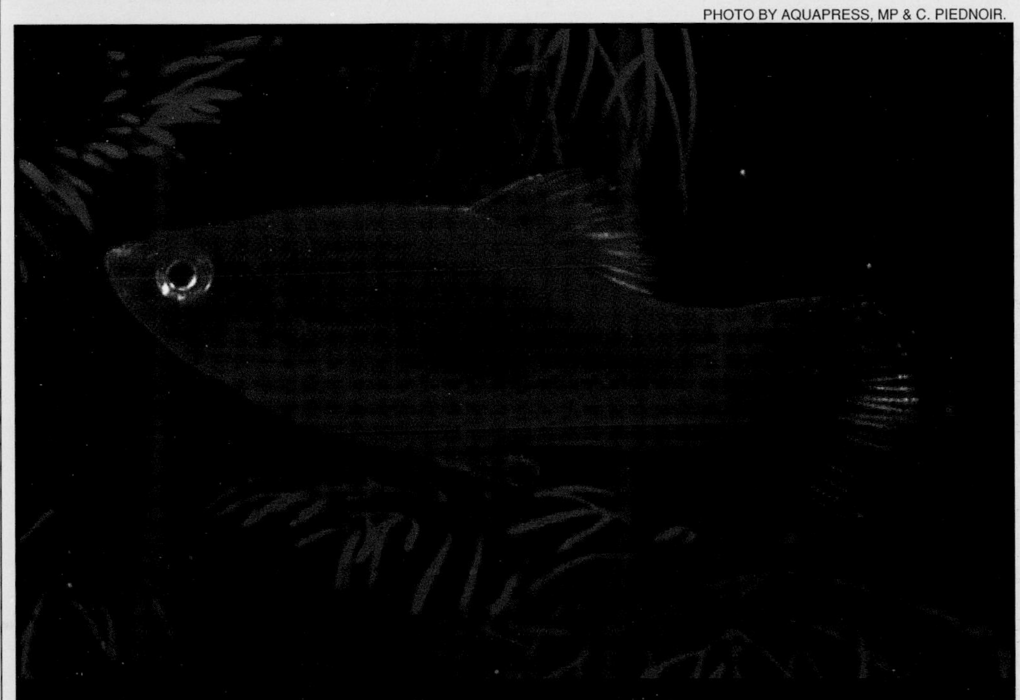

A new strain of blood red semi-wag platy. The black wag pattern is only to be found on the tail and not on the other fins. In the usual wagtail platy all fins are black.

The nitrogen cycle is a delicately balanced process that requires the presence of colonies of certain bacteria to convert ammonia (NH_3) into nitrites (NO_2), and these into nitrates (NO_3). The first two chemicals are exceedingly dangerous to fishes in even very small amounts, the latter far less so. If you were to introduce a number of swordtails and platies into a freshly established aquarium it is very likely most of them would die. The term for this is "new tank syndrome," which means that the water is not yet "mature." By this is meant that there are insufficient beneficial bacteria present to cope with the rate of decomposition. As a result, ammonia and nitrite levels reach lethal proportions.

The fishes may indicate this problem by swimming at the surface gasping for air. You might think that the water lacked oxygen, but it would actually be the damage done by ammonia and nitrite to the gill membranes, thus rendering them incapable of absorbing oxygen. The remedy is to be certain that the water has matured before any fishes are introduced, and then stock the tank at a very slow rate, which is equal to the system's ability to convert the dangerous chemicals into the safer ones.

The denitrifying bacteria are present in the water and all they need is somewhere to colonize. This is the substrate, rocks, filters, indeed, any surface that is well oxygenated.

You can purchase test kits to

measure the ammonia, nitrite, and nitrate levels. When these are virtually nil it can be assumed the nitrogen cycle has been completed and the water has matured. In a newly set up tank these readings will fluctuate wildly, so you must wait until you obtain constant safe readings before fishes are introduced.

You can purchase starter cultures for the nitrogen cycle from pet shops—or you can suspend a sliver of meat in the water for a while. This will also result in colonies of the needed bacteria becoming established. Likewise, young vigorously growing plants will be beneficial in utilizing nitrates and phosphates as food. They will also compete with algae and reduce cloudiness in the water.

FILTRATION

To complete your mini-ecosystem you will need filtration and heating systems. The latter, and lighting as well, are discussed in the next chapter. Filtration is effected by three principle means, mechanical, chemical, and biological.

1. **Mechanical:** This removes non soluble materials suspended in the water by passing the water through a medium, such as nylon, foam, or any comparable inert substance that will block the onward movement of the debris. Such filters also act in a biological manner—as do plants.

2. **Chemical:** This removes many soluble compounds and gases by providing a surface for the adsorption of these. Charcoal is the most commonly used material, but zeolite is also useful. However, the latter cannot be used if the water has been salted as a tonic, or medicine added, because it causes the zeolite to discharge gases, such as ammonia, that might previously have been absorbed. These filters provide mechanical and biological filtration as a secondary benefit.

3. **Biological:** This enables the nitrogen cycle to be completed by provision of a well aerated material, such as gravel, on which beneficial bacteria (nitrobacters) are able to live. Most surfaces are thus biological, but you can also use an undergravel filter system that draws water through the gravel in order to ensure it is especially oxygen rich. Biological filters, such as gravel, effect a mechanical role as well.

Of course, if a filter becomes blocked with debris it will no longer function properly, and in the case of chemical filters they may even start to discharge the dangerous compounds they have neutralized. For this reason all filters must be cleaned or replaced periodically, as they become less effective. Do not clean all filter material at once as this will eliminate all the beneficial bacteria. Always leave enough filtering material to keep things operating until the new or cleaned material has cycled.

AERATION AND FILTER SYSTEMS

The object of aeration is to ensure that a given volume of water can absorb its maximum potential of oxygen. This is

PHOTO BY DR. WARREN E. BURGESS.

This school of poor quality sunburst platies shows regression. Note the black spot appearing on the caudal peduncular area of a few of the fish and the lack of golden luster for which this strain is known.

achieved by passing air from a small pump down a plastic tube and releasing this through one or more porous materials, such as specially made air stones. Most of the rising column of bubbles burst at the surface, thus creating miniature waves. These, in effect, increase the water surface area. It is this, rather than the bubbles releasing oxygen as they travel on their way to the surface, that increases the oxygen content. As a side benefit, the rising column of bubbles draw water behind them, thus helping to create circulation: this helps to avoid temperature stratification.

If a filter system is a feature of your set up, which it should be, separate aeration is not needed. This is because the filter system works on the basis that it takes water from the lower levels of the tank (where it is dirtiest and least oxygenated). It is then passed through the filter(s) before being returned to the tank at its surface, either as a single stream, or via spray bars. Both effectively agitate the surface. In addition, as the water passes through the air it will be increasing its oxygen content before it even touches the surface.

Filter systems are, of course, even more effective than simple aeration in creating circulation, because they move larger volumes of water. You can select from a very wide range of filter systems that include those that are operated via an air lift (aerator) through very sophisticated power driven units. These may be internal or external to your aquarium.

Truly beautiful young sunburst platies.

PHOTO BY DR. WARREN E. BURGESS.

PHOTO BY DR. HERBERT R. AXELROD.

This is a male golden red maculatus platy.

It is not necessary to have a filter system that is creating a mini Niagara Falls in the amount of water it draws and returns to the tank. Indeed, if the filter moves too much water it may impede biological action and be too turbulent for good plant growth. Your pet or aquatic dealer will advise you on a system suited to the size of aquarium you are thinking of purchasing.

Do bear in mind that internal filter canisters take up quite a lot of room, so in the medium to small aquarium it is best to have an external filter. This is more easily serviced as well. Once you have a filter system in operation, do not turn it off other than in emergencies or for cleaning. If you do, it is likely the nitrobacteria population will soon start to die (due to lack of oxygen), with the resulting excessive pollution in the water.

It may seem to the first time aquarist that there is a tremendous amount of knowledge needed to maintain a healthy aquarium, but this is not so in reality. Once you have the aquarium up and running it is merely a case of periodic checks that the aspects discussed in this chapter are within safe tolerances. Such checks will become routine to you, and you will quickly be able to detect when things are not as they should be.

A final comment in preparing an ecosystem is that no matter how good the filter system is there is always the possibility that "unknown" substances could build up in the water that are not being removed by filtration. To eliminate the risk that this might be so it is wise to periodically replace 20-25% of the water in order to continually dilute such compounds. This should be done about every 2-3 weeks.

THE AQUARIUM AND ITS APPLIANCES

Before you go about the business of selecting an aquarium and its ancillary appliances, you should give careful thought to its size. Very often beginners purchase a unit that proves far too small for the number of fishes they envisage keeping. This is invariably because they obtain an aquarium before they find out about the number of fishes the tank will hold, or because finances dictate the smaller unit. Let us therefore commence by reviewing the advantages of the larger tank, and why it is better to save some extra cash in order to purchase one.

First of all we should establish what is and what is not a "larger" aquarium. Tanks that have less than a 12-gallon capacity are small. They will have dimensions of 62 x 31 x 31cm (24 x 12 x 12 in) or less. A medium tank will measure about 62 x 45 x 38cm (24 x 18 x 15 in) and holds 20 gallons of water. A large tank will exceed this and have a capacity of 23 or more gallons. Its dimensions will be in the order of 93 x 38 x 31cm (36 x 15 x 12). This is a good size to have: it is spacious without being too large and costly.

The advantages of a large unit over smaller ones are as follows:

1. It will accommodate more fishes—enough to make an impressive display.

2. It has greater scope for you to aquascape (furnish) it without things becoming crowded.

3. It is more esthetically pleasing because it is more obvious and impressive than a small unit that might be hardly noticed.

4. It will retain its temperature more readily than the smaller unit.

5. It will maintain its water quality more readily. Put another way, this will not deteriorate as rapidly if for one reason or another you have to miss a regular cleaning chore.

You will also find that a larger tank will probably be constructed of better materials than many of the smaller units, and the plastic ones will not scratch so easily, or become "yellow" as they age. Of course, if you plan to breed your fishes smaller units can be used, but we are thinking here in terms of a main display aquarium.

STOCKING LEVELS

To fully appreciate the size you will need you should think first of how many fishes you would like to keep, then see how large the tank would have to be based on the following calculations. These are for a tank that has no mechanical aids (aeration or filtration) that would dramatically increase the stocking level. However, always remember that the more fishes

PHOTO BY DR. HERBERT R. AXELROD.

Two male hifin or sailfin variatus platy males. Only the males have the elongated dorsal fins.

kept beyond the basic stocking level, the greater the risk of problems in the event of mechanical or power failure. Also, in a crowded aquarium, if a health problem should arise, this will rapidly spread through the population. The generally accepted formula for calculating tropical fish stocking levels is to allow 75 cm^2 (12 in^2) of water surface area for each 2.5 cm (1 in) of fish body, excluding the tail. This means that tropical fishes can be stocked at double the amount recommended for coldwater species. The depth has little influence on matters, but it is useful in that it gives the fishes more swimming room. In a community tank this can be utilized to feature species that swim at different levels, e.g. surface swimmers, mid-level fishes, and bottom swimmers.

The surface area is determined by multiplying the length of the tank by its width. This is divided by 75, and this multiplied by 2.5 to give the total number of fish-body centimeters. The average length of the fish is then divided into this to yield the number of fishes. It follows that if calculations are based on youngsters the tank will hold many more of them than if they are adults, so allowance must be made for this reality. Unless you have two or more tanks available to house the growing stock, it is best to use adult length as the basis for stocking levels.

A selection if potential stock levels is as follows, based on average platy adults.

TANK SIZE Length/Depth/Width	Surface Area	Gals US/UK	Total Body Length	Number Of Fish
31x25x25 cm 12"x10"x10"	775 cm^2	5/4.2	26 cm	3.9
46x31x25 cm 18"x12'x10"	1,150 cm^2	9.4/7.8	38 cm	5.7
62x38x38 cm 24"x15"x15"	2,356 cm^2	24/20	78 cm	11.7
93x38x38 cm 36"x15"x15"	3,5344 cm^2	35/29	118 cm	17.6

surface areas, you will appreciate that tall "designer" tanks are far less suitable for fishes than are the standard rectangular units, and can hold fewer fishes. With a good aerating filter system you can double the basic stock level, but more than doubling it is moving toward less satisfactory husbandry techniques, even though it can be achieved.

WATER VOLUME AND WEIGHT

Knowing the volume and water weight of your aquarium has two advantages. This information will be needed (1) if medicines are ever added to the water, and (2) to establish the weight of the aquarium when deciding if a given support is sufficient to bear this weight. Water is deceptively heavy.

The volume is established by multiplying the length x depth x width in the required

Beautiful original redtail variatus platies. These are almost the wild colors. Fish like these were inbred for interesting color variations and eventually all the variatus strains were produced from original fish like these.

PHOTO BY HANS JOACHIM RICHTER.

DRAWING BY JOHN QUINN.

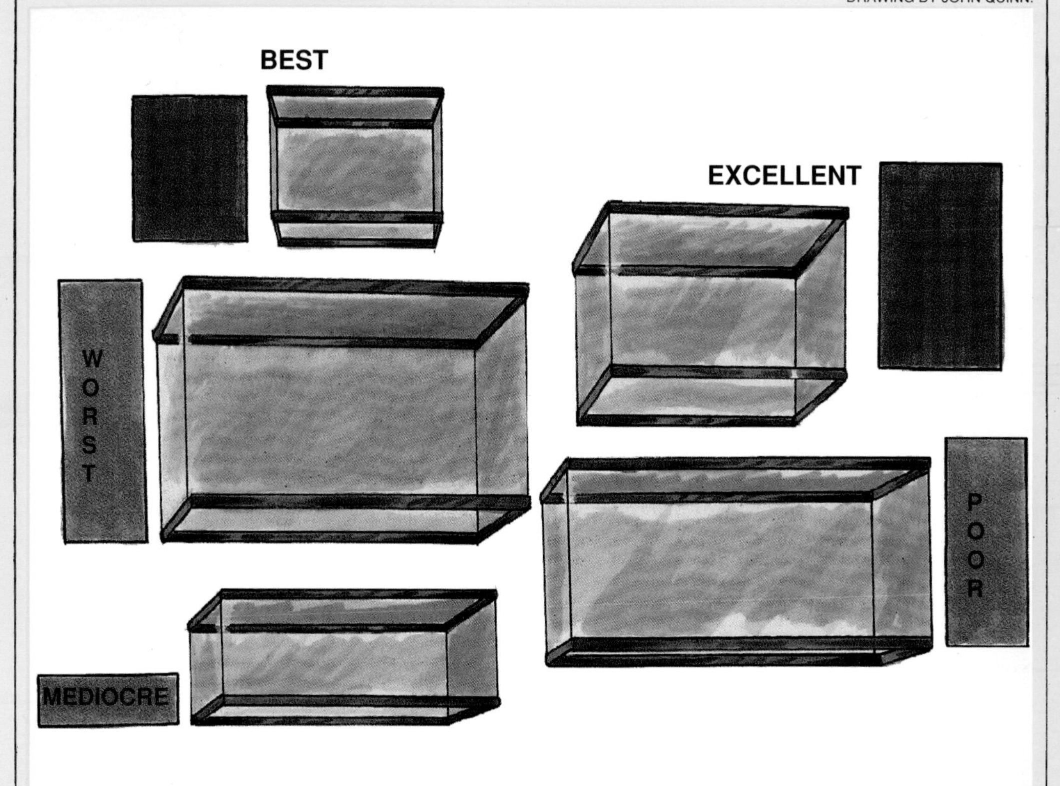

The best fish tank is the one with the most surface to air contact. In arithmetic terms, the closer to a circle in cross section, the better. The BEST tank is a cube, with three equal dimensions. The worst is a long, skinny, deep tank.

measurement, then dividing this by the units per gallon in that measurement. For example, if an aquarium is 62 x 45 x 38 cm (24 x 18 x 15 in) its volume is 106,020 cm^3 or 6,480 in^3. Since there are 3,785 cm^3 in a US gallon, the tank holds 28 US gallons. There are 231 in^3 in a US gallon, so using this measurement the volume is again 28 US gallons. A few conversions will also be useful to you so that you can change to whatever measurements you wish.

US gals to UK gals multiply by 0.833

UK gals to US gals multiply by 1.2

US gals to liters multiply by 3.8

UK gals to liters multiply by 4.5

Cu cm to liters divide by 1,000

1 US gallon weighs 8.35 lb = 3.8 kg

1 UK gallon weighs 10 lb = 4.5 kg

The 28-gallon tank in our example would have a weight of 234 lb—which is equal to that of a very heavy man, a sobering thought if you assumed the tank could be placed on a less than very well supported shelf! A 35-gallon tank would weigh 292 lb, plus of course the weight of the tank itself. The rocks and gravel can be ignored because their weight would be more or less equal to the water they displaced.

HEATING

There are a number of ways in which your aquarium can be heated— such as space heating, heater pads under the aquarium, and gravel heating, but by far the most popular is a combined heater-thermostat unit placed into the tank water. There are two types, one that is clipped to the side of the tank with its controls above water level, the other is fully submersible. The latter are perhaps your best choice, and the more popular.

The down side of the combined heater-thermostat is that if one element fails (heater or thermostat) you must replace the entire unit. However, the well established name brands are very reliable these days. The best ones will control the temperature within 0.25 of a degree. Select one that has a pilot light to indicate when it is working, and that can be adjusted to whatever temperature you wish. Be sure the heater is submersed in a vertical position and does not rest on the gravel, otherwise its performance can be greatly diminished. The glass may even shatter.

You can purchase a plastic perforated heater sheath to remove the risk that your fish might accidentally rub against the unit and burn their skin. In a large aquarium it is prudent to use two heaters rather than one of greater power. The reasons are twofold. Principally, if one should fail the other will probably provide sufficient heat to keep things going until the failed unit is replaced. Secondly, if a powerful heater is used that can take the water to the desired heat it will probably be capable of taking the temperature beyond the safety level should the thermostat malfunction— you might cook your fishes if you didn't notice the escalating temperature! With this fact in mind it is always wise to feature at least one thermometer (external or internal, there are many to choose from) in your set up so that you can make a habit of checking this every time you view your display.

Deciding on the wattage of a heater must take into account a number of factors:

1. The differential between the ambient temperature and the water temperature required.
2. The volume to be heated (large volumes have less heat loss than small ones).
3. Whether a canopy and glass cover are fitted. If they are, the water will retain its heat better.
4. The site of the aquarium (to be discussed in the next chapter).

Taking an average situation the following should prove adequate:

15-40 gallon units = 5 watts per gallon

40+ gallon units = 3.5 watts per gallon

PHOTO BY DR. HERBERT R. AXELROD.

The female golden wagtail on the top was bred with the reddish gold male in the center to produce the wagtail bleeding heart.

Heaters are sold in wattage increments of 25 and 50 (75 w,100 w, etc.) so select one as near to the total wattage you need. If 200 watts are required purchase 2 x 100 watts. If the heater is too powerful it will be constantly switching on and off. If it is at its limits it will be on all the time—any drop in the room temperature might mean it may be unable to maintain the desired heat.

It is wise to precheck that your heater is performing adequately before fishes are introduced. You should also stock a spare heater so no time is lost in replacing one that fails.

LIGHTING

Lighting is important in your aquarium for three main reasons:

1. It is essential for photosynthesis to take place. Photosynthesis is the process by which plants absorb carbon dioxide and utilize the energy from light to grow. It creates chlorophyll, the green pigment found in plant cells.

2. It is needed by the fishes to enable their body rhythms to function correctly. It also has a strong influence on their reproductive capability.

3. It is essential for your own viewing pleasure. This is because without sufficient artificial light

Very rare fishes are the black platies. One, the female, is all black while the other, the male, has red in the unpaired fins.

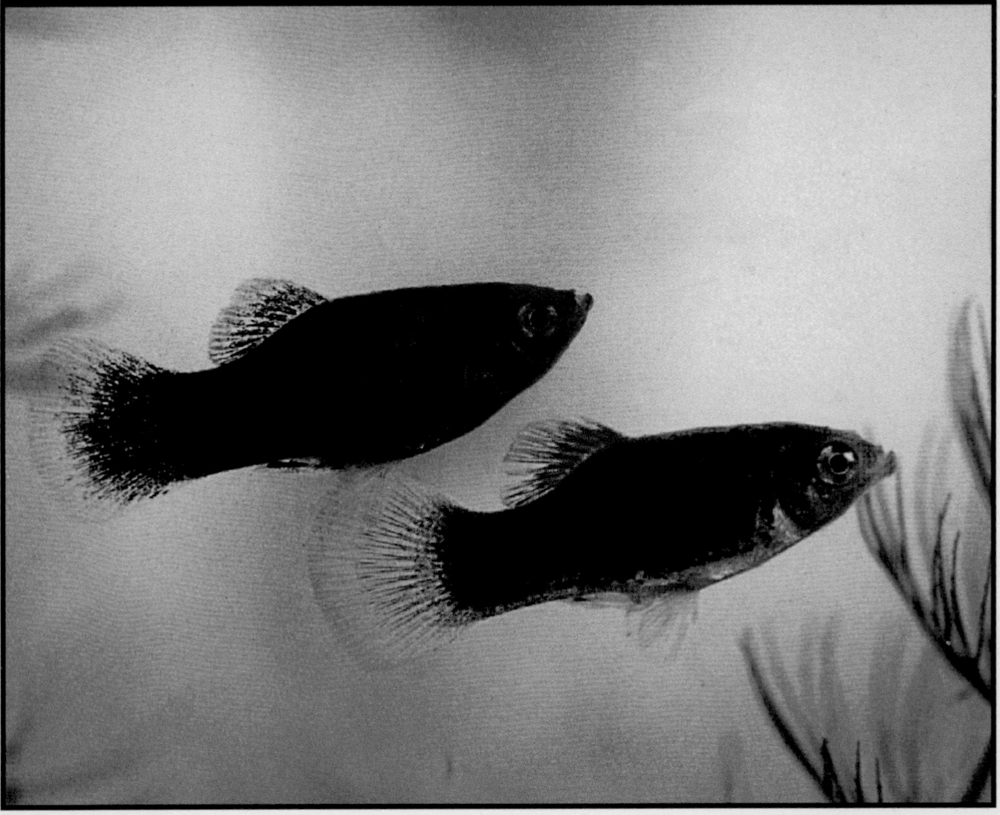

PHOTO BY DR. HERBERT R. AXELROD.

PHOTO BY DR. HERBERT R. AXELROD.

The golden comet platy is so named because of the streaks in the outer margins of the tail fin.

you simply would not be able to see into the aquarium very well.

In addition, if the only available light came from the front or sides of the aquarium, the plants would grow toward the light source, often taking unnatural positions. Some fishes may also swim at an angle because they, too, are phototropic, preferring to have their backs angled toward the light source. Finally, without good illumination much of the beautiful swordtail and platy colors would not show themselves to good effect.

Selecting Lights: Although you have numerous lighting options (including various spotlights and tungsten lamps) by far the most popular are fluorescent tubes. These are cool operating, economical, and can be fitted under the canopy of your aquarium. There is an extensive range of wave lengths to choose from, including daylight and those with a bias toward one or the other end of the spectrum (blue, orange, or red) to encourage plant growth.

If your hood will accommodate two tubes it is best to feature one tube from each of these types. This will provide good color illumination of the fishes and the special rays beneficial to the plants. Be aware that color biased lights will modify your fish's colors when compared to natural daylight tubes. Some experimentation may be required to find the ones that you, the plants, and the fishes like best.

Photoperiod: Tropical fishes live where there is approximately equal periods of day and night, so a 12 hour on/12 hour off schedule is recommended. This can be modified to about 11 on/ 13 off if plant growth is too rampant. In order to avoid "night-fright" when lights are suddenly switched off-—or "light-fright" when suddenly switched on, switch the tank lights off a short time before the room lights go off. Alternatively, a low wattage night light can be put on before the tank lights are switched off.

Do check what the lighting needs of proposed plants should be because vegetation can be more particular than the fishes with respect to this and, of course, heating requirements. It is best to commence with well established hardy plants before trying out the more delicate species.

Wattage Needed: As a guide, you can start out by working on the basis that your tank will need 10 watts for every 31cm (12in) of its length, or, if a tungsten bulb is used, 40 watts for the same length. If your tank is unusually wide you can calculate on the basis of 10 watts for every 900 cm^2 of surface area. Round up or down to the nearest wattage that is available in a length that will fit under your canopy.

Always fit a cover glass over the tank if a guard is not already a feature of the aquarium. This will prevent condensation from reducing the effective life of the electrical fittings. Keep the glass clean, otherwise much of the light's benefit to plants will be lost. Even if this is done, your tubes will lose about 20% of their efficiency over the course of a year. It is best to routinely replace them, even if they are still working.

OTHER ACCESSORIES

There are a number of accessories you can purchase that will be needed for routine cleaning and general maintenance. Most are not costly.

1. **Glass scrapers:** These are for removing excess algae from the side and viewing glass. A little algae is beneficial to the fishes, but excess algae looks unsightly. This indicates that there is too much light, or not enough competitive plants.

2. **Gravel cleaner:** This is a siphon tube that contains perforations large enough to allow debris to pass through, but not the gravel. You move it around the gravel vacuuming one area at a time.

3. **Length of smaller diameter siphon tube**. This is always useful for getting into those areas that the gravel siphon is too large for.

4. **Aquarium nets:** One or two small nets are handy to scoop out suspended debris, as well as to herd fishes toward containers for removing or inspecting them. It is always wise to avoid netting the fishes themselves because of the potential damage that can be done to their scales and fins.

5. **A 5-gallon aquarium** will be useful as a hospital tank, for quarantining new arrivals for the display tank, or for getting new young plants established.

PHOTO BY HANS JOACHIM RICHTER.

A beautiful golden tuxedo platy female.

6. **A good hand lens** is very useful for inspecting fishes if external parasites are suspected.

Regular partial changes of water in any aquarium, especially a platy aquarium, are of great benefit to the fish. Such changes can be made with siphons and buckets, or, in a labor-savings manner through the use of a water-changer. No aquarist can be successful without this device. Photo courtesy of Aquarium Products.

7. If you do not want to make your partial water changes manually there are **automatic water changer** appliances available, but they are not essential. Buckets and jugs are still very convenient, so have one or two just for your hobby use.

8. **Planters:** These long probes are very useful for inserting additional plants into the aquarium without the need for disturbing things with your outstretched arm—which is never advised if electrical appliances in the aquarium are switched on.

9. **Plant root weights** and small plant tubs. These will be invaluable for establishing young plants.

As with any hobby, the largest cost is in the initial setting up. Maintaining an aquarium, however, is very inexpensive compared to most other hobbies. You will find that it is certainly worthwhile to have the items discussed in this chapter, plus aquarium decorations and plants, from the outset. It is not necessary, nor even advisable, to invest in the fishes until you have given your aquarium time to mature.

SETTING UP THE AQUARIUM

Before you actually go out and purchase your aquarium it would be prudent to consider exactly where it is going to be placed. You should also make some sketches of your proposed aquatic scene so that you can purchase the right decorations to achieve the desired effect. Finally, allow plenty of time for the actual setting up process, as it invariably takes longer than might be thought.

POSITIONING THE AQUARIUM

The three considerations in positioning the aquarium are heat, light, and its weight. It should not be placed any place where the surrounding conditions might create rapid fluctuations in the water temperature. For example, a tank located near a window can rapidly heat up during warm weather. The direct sunlight would also dramatically increase the rate of algal growth. Another location that is not desirable is over, or next to, a central or other heating or air conditioning unit. With heat loss as a factor do not place it near, or opposite, a door that leads to an area where the temperature may be much lower.

The preferred location will be a point in your living area where the temperature is as constant as it will get. At the same time the site should be such that when seated you have the best view of the display. With regard to the support, this must be solid enough to take the weight you have calculated for the tank when it is full of water—and then some for a safety margin. It is not unknown for a large tank to actually crash through the floor of a home! It may not do so immediately, but it can happen due to the continual stress of the weight. Be very satisfied as to the strength of the support for the tank itself—and that which is supporting it.

The supporting surface must be very level, otherwise there is the risk that the water pressure on any unsupported area of the glass base could cause this to be stressed to the breaking point. Very minor deviations in the level can be overcome by placing cork or polystyrene tiles under the tank—these will also be beneficial by having an insulating effect. The sides, and indeed the back panel of the aquarium, can also be insulated with cork to minimize heat loss. This may not be needed in a centrally heated home, but could well be worthwhile if the room is not so heated, thus likely to experience a significant overnight drop in temperature.

AQUASCAPING

Your finished display is called the aqua scene. This is

PHOTO BY BURKHARD KAHL.

Red iridescent Mickey Mouse platies. They are also called *golden red* platies.

determined by what is placed into the aquarium, and what is immediately surrounding it. If the latter is not to be an integral part of the scene, which we will discuss later, it should be neutral so that it does not detract from the aqua scene. The major components of an aquarium are as follows:

Rocks and Gravel: It is always best to use rocks that have a neutral effect on the water. These include granite, slate, and basalt. Avoid any with streaks of metal in them, or those that are calciferous, such as marble, chalk, or dolomite. Likewise, oyster shell is unsuitable to use in the substrate. Rocks that are well rounded are best so that they give the appearance of being well worn.

The gravel should be of a medium to dark color. This makes the fishes more at ease and mirrors their natural environment. The gravel size should be neither too small nor too large. In the former case it packs down and restricts the flow of oxygenated water to the roots, and thus makes life more difficult for the plants, and at the same time reduces the efficiency of the biological filtration. If too large, uneaten food can accumulate in the little crevices.

Wood: Driftwood and bog wood are fine for the aquarium because any potentially toxic material will have been leached out. They make very natural "props." Bamboo and other floating woods can be decorative, but will need to be weighted down with rocks or gravel.

Artificial Decorations: The range of commercially manufactured rocks, woods, ornaments, and plants is quite breathtaking these days. The better quality examples look just like the real thing and can be used with great effect in conjunction with real ones. There are of course all manner of novelty items, such as treasure chests, divers, galleons, mermaids, and their like. The purist aquarist shrinks at the sight of these, but other hobbyists think they are great, and really enjoy planning a novelty tank with the utmost care. It's your hobby, so go with what appeals to you and your family.

Plants: There is a very extensive range of tropical aquatic plants from which you can choose. Your local pet shop, and especially your aquatic dealer, will stock a selection of the popular and hardy varieties. These are the ones to choose from. Try and obtain young examples. These will establish themselves better under your aquarium conditions than if mature specimens are transplanted. These may have been grown under conditions rather different from the properties of your water.

Your selection should include

Aquarium ornaments serve many purposes. While their primary purpose is making the tank look beautiful and interesting, they also serve as a hideout for small fishes and a base upon which algae can grow and be used as food for foraging platies. Photo courtesy of Blue Ribbon Pet Products.

There are many different kinds of shapes and types of ornaments. An airstone placed within, behind or under an ornament makes it very interesting.

plants of differing heights for the foreground, middle, and background. It is better to try and create an impressive display using a few species, rather than attempt to include too many varieties. Purchase a solution of plant disinfectant so that you can wash the plants before they are placed into your display unit. This minimizes the risk that pathogens might be introduced into your tank via the vegetation.

EXTERNAL AQUASCAPING

Rather than be looking at the wallpaper behind your aquarium when you view your display, you can do much to make things far more interesting. There are two options:

1. Your aquatic dealer will stock either flat or three dimensional mural scenes that can be fitted to the back and side panels of your aquarium. Alternatively, you could place cork on these panels. This is attractive and acts as a heat insulator as well.

2. You can create your own diorama by placing rocks, wood, and other decorations on a "stage" at the back of the aquarium. You can also illuminate this scene to give it even more impact. This will give the viewer the illusion that the aquarium is much larger. At the same time you can change the scene very easily from time to time. It can include materials, such as calciferous rocks, that would not be advisable to place in

the aquarium itself. For the best effect you must of course keep the back panel clean, and leave sufficient space between the plants so that the scene can be seen. An external diorama can be set up before or after the display tank is attended to.

SETTING UP THE DISPLAY

Before you actually begin setting up your display it is best to wash and rinse everything (including the tank itself) that will be going into the water in order to remove dust from it. This is especially applicable to rocks, gravel, and wood. These can be immersed for a few minutes in a solution of household bleach to kill unwanted microorganisms. Be sure to thoroughly (!) rinse any item treated with bleach.

You can prepare a few buckets of water the day before so that the chlorine content has time to dissipate—helped by your occasional stirring. Leave the water in your living room so it will reach room temperature by the time it is needed.

The actual setting up should follow a predetermined plan. The following is suggested:

1. If an undergravel filter is to be used this should go in first. Make sure it is a tight fit. Any small gaps can be filled with aquarium silicon sealant so that water will not by-pass the main filter bed en route to the uplift tube.

2. A layer of gravel should cover the undergravel filter or aquarium base. Into this the larger rocks and driftwood can be inserted and gently pressed down so that they leave no pockets where uneaten food and debris might collect.

3. You can now build up the gravel so that there is a natural slope from back to front—7.5 cm to 2.5 cm (3 in to 1 in). Rocks and plastic or Plexiglas (Perspex) strips can be used as retaining walls if terraces are required anywhere. Bond these to a base so that the weight of the gravel will not move them out of place.

4. At this stage you can gently pour some of the aged water into the tank. The water should drop onto a saucer or cardboard so that it does not unduly disturb the gravel. You need enough to allow plant leaves to float so that they are not in your way as you insert the roots.

5. Insert plants and place some fertilizer around them. Cover small plant tubs with pebbles so that they cannot be uprooted by the fishes.

6. With the gravel and all the rocks in position, you can now insert the heater and the canister filter if these have been chosen. If not, the external filter tubes can be fitted, as well as the air tubes and their outlet stones. These can be discretely hidden behind rocks and plants

7. With all the major components now in place you can finalize your decisions regarding minor decorations before filling the aquarium to about 5 cm (2 in) below its capacity. You can now also make your first readings of the pH, hardness, and other properties if the water was not prepared to a particular quality beforehand.

A nice school of sunburst or sunset platies. All have almost the same coloration. Males have more red

8. Before fitting the cover glass and canopy on the aquarium it is suggested that you plug in the heater and see that this is working if it was not previously tested in a bucket of water. If it was, attend to these final details and switch everything on. Never (!) place your hands into the water when electrical appliances are plugged into the sockets.

9. Once the water has been adjusted to the needed temperature you can add the biological filter starter culture. You will then go into a "wait" mode.

Each day you can make water quality tests, which will give you good experience running such tests. Only when the water has matured and reached the desired readings should the fishes be introduced. These should be inexpensive fishes. If these prosper you can then add more stock a few at a time—never overload your filter system with the addition of too many fishes at any one time.

During this initial period keep an eye on the plants. Remove any that die, or any leaves that have clearly departed the living world.

INTRODUCING THE FISHES

The fishes will be transported home in a plastic bag. This bag should be floated on the water surface to allow its temperature to equate with that of the tank. Next, open the neck of the bag every once in a while and allow a partial exchange of water so that the new arrivals can become acquainted with the properties of their new water gradually. Finally, gently open the neck again and let the fishes swim into their new environment. This whole process will take about one hour. This method of transferring fishes from one tank to another should also be used when you start to breed your own fishes.

FISH NUTRITION

Platies are very easy to cater for with respect to their nutritional needs. Scientifically, they are known as omnivores, which means that their diet is made up of both plants and animals. They show a bias toward being herbivorous (plant eaters) when mature, but are more carnivorous (flesh eaters) when juvenile. Before discussing particular foods, and the forms they are available in, a rather important general observation that is invariably appropriate to the first time hobbyist is worth mentioning.

OVERFEEDING

Unlike many other popular pets, whose diets are almost totally controlled by the owner, this is not as true in a well established aquarium. This is a mini ecosystem in itself. Apart from many microorganisms, it will also contain various algae and other plant forms. Your fish will browse on these, as well as consume the minuscule life forms that you may not even be able to see.

As a result, it is very easy to overfeed your fishes. It happens that obesity is not a major(!) problem in fishes whose body temperature is regulated by the environment, so they have little need for excess layers of insulating fat. The more dangerous consequence of overfeeding is that the uneaten food will decompose in the water. This places an unnecessary strain on the nitrogen cycle, which may not be fully completed, thus increasing the risk of dangerously high nitrite levels.

The fact that the daily nutritional needs of even a small collection of fishes appears to be no more than a small quantity in your hand, creates the strong temptation to supply too much food. Your first feeding rule should be to only supply that quantity which is readily devoured by your fishes as it sinks to the substrate. Once you notice disinterest in the food you can then estimate how much to reduce the amount (if necessary) at the next feeding. The other aspect of feeding these fishes is that herbivorous animals prefer to eat a little at a time and often, rather than gorging themselves on their food like many carnivores. This fact underscores the comment about not feeding too much at any one meal. Always take the time to observe your fishes when they are feeding. By so doing you will soon know how much to supply. You will also be aware of the feeding habits of the individual fish. Should these habits change, this may be the first indication of a health problem.

COMMERCIAL FISH FOODS

Foods are made up of proteins, fats, carbohydrates, vitamins, minerals, unidentified substances, and water in an infinite range of ratios of one to the other. When these are prepared by commercial manufacturers, preservatives and color dyes are added to make them more appealing to us—and via us to our pets. Likewise, packaging is also carefully prepared to entice us into buying this or that product, regardless of whether the product is actually any good!

You should not, therefore, be persuaded to purchase foods based on their packaging, or on the fact that the foods appear to have a beautiful array of colors (which of course are totally artificial). Instead, take more note of what the small print on the label states with respect to the percentages of its composition. If there is only a brief or no content list it would be wise to ignore that particular food. Fishes need a dry matter protein content in the range of 45-50%. It may be a little lower for adult non breeding fishes, and rather higher for growing juveniles. A great deal of research has been made into the nutritional needs of fishes. This has resulted in an excellent range of commercial foods that are wholesome, and which are often fortified with extra vitamins. Many are prepared with particular

PHOTO BY BURKHARD KAHL.

Red dorsaled golden Mickey Mouse platies.

feeding groups in mind (carnivores or herbivores) and at a size suited to specific species, or groups of these.

There are foods for fry, juveniles, and adults, as well as ingredients that are drawn from a vast number of sources—cereal crops, seeds, meat, fishes, and invertebrates. These foods are very convenient to use if stored in cool dark cupboards, or the refrigerator, according to their type. They are available in the following forms:

Flakes: These may be fast or slow sinking. You need the slow type to meet the needs of platies, which are basically surface or upper water feeders. Quality flakes will not cloud the water.

Tablets & Cubes: These can often be stuck onto the front or side panels of the aquarium to form feeding stations that allow you to more readily observe individual feeding habits.

Powders & Liquids: These are excellent for fry as they are obviously very small particles. They will find their way to the areas of the tank where youngsters like to take refuge among plants.

Freeze Dried: These come as individual foods, such as shrimp, tubifex worms, *Daphnia*, indeed most popular invertebrates.

Deep Frozen: Once again just about all live foods, as well as various algae, can be obtained in this form. It enables you to feed high protein complete foods without the need to gather these from the wild. In the latter case there is a much greater risk that pathogens might be introduced into the aquarium via the food organisms.

Commercial Live Foods: If you prefer to include some living foods in the diet of your fishes these are best obtained from commercial sources to minimize the risk of attendant parasites or pathogenic bacteria. All common freshwater and marine invertebrates, together with many land creatures—insects and their pupae, and various worms—are available from your aquatic dealer, or direct from commercial breeders.

The major advantage of live foods is that any that are uneaten will not pollute the water, but swim around (if they are freshwater species) to be consumed once the fishes get hungry again. Alternatively, certain of these foods, such as worms, can be placed into special holders (available at your pet shop). These enable the worms to "escape" over a period of time, and the fish will quickly devour them before they have actually left the container.

Live foods are especially beneficial to breeding stock and to their fry, both of which seem to be more stimulated and satisfied at this important time of their lives.

NON COMMERCIAL FOODS

Just about all the foods that you eat can be given in appropriately small amounts to your fishes (but never sweet, stodgy items). This keeps the food costs down and adds variety to their diet. You can either feed them as individual items or make a one time mixed mash of them and store this in your refrigerator (in the freezer compartment) in ice cube holders. This way, you can take out a cube at a time, thus avoiding the need to thaw and refreeze large portions, which is not advisable.

The mix can be held together by utilizing a binding food, such as porridge. Of the many items your fishes will enjoy, the following is but a small selection: Garden peas, carrots, spinach and other green vegetables, fruits, potatoes, cheese, bacon, ham, meat, poultry, fish, any crustacean foods (such as shrimp, crab, lobster), bread crumbs, and commercial dog and cat foods (be these canned or in biscuit form). Avoid oily or fatty foods.

Initially, your fishes may not show any special liking for these foods, largely because they may never before have received them in these forms. But if you persevere, and note which are taken or attract the interest of the fishes, you will find that many will eventually become greedily taken by your fishes. During this familiarization process it is important that only very small amounts are given in order to reduce the risk of polluting the water.

VACATION FEEDING

When you are away on vacation the feeding of your fishes can be a problem. This is because a friend or neighbor unfamiliar with the small amounts needed per meal will invariably overfeed—with the resulting pollution of the water. If your fishes are in very good health, if you will be away for one week or less, and if the aquarium is of good size and well planted, it may be better not to have an inexperienced person feed them at all. Most fishes can survive such a period without undue problems.

Alternatively, place minimal rations in a number of sachets and tell the helper to give one sachet per day. Explain the dangers of pollution to them, and that excess food might kill your fishes. This should make them wary of throwing an extra handful of flakes into the tank!

However, even if you do not plan to have the fishes fed you should make arrangements with someone to visit your home each day to see that all the equipment is working. Leave them the name and telephone number of an aquarist friend, or your local dealer, who can be contacted if there are any problems—advise the aquarist or dealer of the situation. If you will be away for more than one week, some dealers make service visits (including feeding every few days), and the cost is well worth it if you have an expensive display.

To conclude this chapter, the two golden rules of feeding are again stressed. The first is to feed small amounts, and feed the fishes a number of times per day. The second is to provide a broad and varied diet. Never allow the diet to become boring by feeding the same items continually.

SPECIES OF PLATIES

In Chapter 1 the method of scientific classification of fishes was discussed. Here we will look at various platy species. This done, we can then look at examples of popular varieties that have been created by manipulating the gene mutations that always appear in animals when they are selectively bred under domestic conditions.

One of the problems that results when mutated forms are hybridized with different species, is that it becomes virtually impossible to identify what species the resulting fish actually are. This is compounded by the fact that hybrids of virtually all swordtail and platy species, and other species in the family Poeciliidae, are possible. In a nutshell, the consequence is a total mess from the purist's viewpoint. This is the current situation in these particular fishes.

When you purchase your platies they may have a species label appended to them, but this should be viewed with a degree of doubt. Most commercial stock is the result of much mixed breeding. As far as their actual scientific standing is concerned, the situation is only marginally less confusing. Different taxonomists have at one time or another raised or lowered the status of a given fish to or from the rank of species or subspecies. Throughout the history of these fishes they have also placed them in different genera, and given what are now considered the same species different trivial (specific) names. What may be no more than variants of a subspecies have been appended species and subspecies names. The end result can and does create only confusion in the mind of the hobbyist.

***Xiphophorus maculatus* (Günther,1866). Common Platy or Moonfish**

Males about 3 cm (1.1 in), the females to 6 cm (2.4 in). The species is found from southern

A pair of wild *Xiphophorus maculatus* showing the tuxedo pattern in the male. This strain is from the Rio Papaloapan.

PHOTO BY LOTHAR WISCHNATH.

Mexico to Honduras in slow to fast moving lowland streams. The coloration is extremely variable. Green-brown on dorsal surface becoming bluish on the flanks, with ventral part very pale. Black body specks are typical, and the peduncle contains one or more much larger dark blotches. Some orange-red may be apparent.

The body is deeper than in the swordtail. There is no sword in the males. The crescent shaped dark marking in the caudal fin gives rise to the common name of moonfish. Care is as for the swordtails, though platies may cope somewhat better with slightly lower temperatures than will swordtails. Temperature range is 64-77°F (18-25°C), with the optimum somewhere between. These are very amenable little fish that make excellent community or specialist aquarium residents.

Xiphophorus variatus (Meek, 1904). Sunset or Variegated Platy or Platy Variatus

Males to 6 cm (2.4 in), females to 7.5 cm (3 in). Distributed in Mexico. As the name suggests its colors are very variable. The head region of the male is yellow, but there may also be some red. The latter is seen in the tail but this may range to yellow. Flanks are blue-green. Dark flecks are seen on the body, and there are a couple of blotches, or a solid band, on the tail. Dorsal fin yellowish edged in black. Vertical flank stripes are sometimes seen. The female is a paler version of the male, which has no sword. The related X.

ALL PHOTOS BY LOTHAR WISCHNATH.

Xiphophorus maculatus male from the Rio Papaloapan.

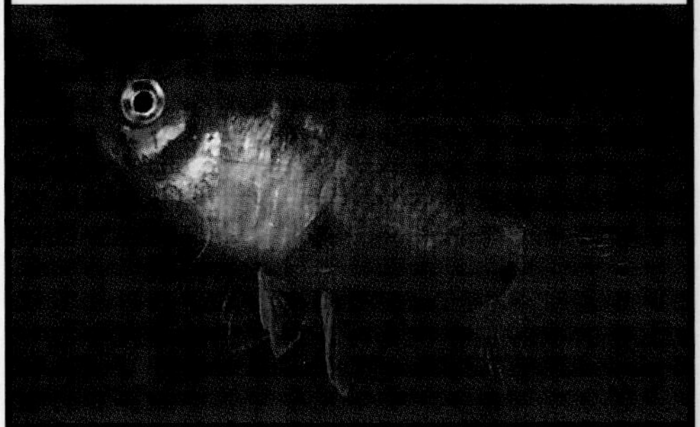

Xiphophorus maculatus female from the Rio Jamapa.

Xiphophorus maculatus male from Belize.

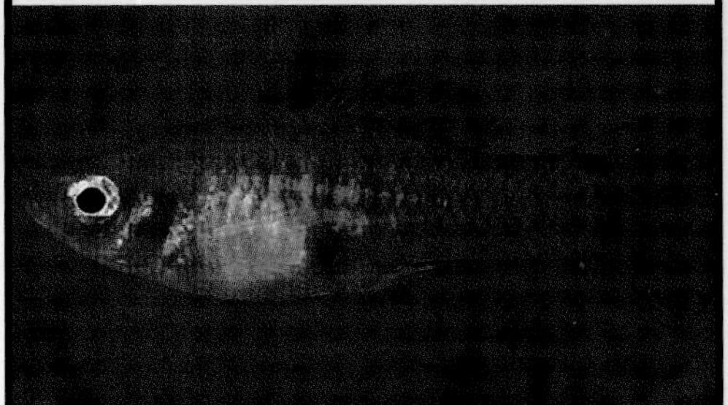

Xiphophorus variatus male from the Rio Mante.

evelynae is paler than *X. variatus*, and usually is well marked with vertical stripes on the flanks. Most of the popular platy variants were developed from this species and *X. maculatus*. Temperature range is 59-77°F (15-25°C), with the optimum being about 70°F (21°C).

Xiphophorus couchianus (Girard, 1859). Monterrey Platy

This species inhabits northeastern Mexico and is now very rare, being classified as endangered. Size ranges from 3.5 cm (1.2 in) in males to 6 cm (2.4 in) in females. It is rather drab colored, being shades of brown and green-brown with a hint of red-orange on the flanks. The under body has a swollen appearance rather like a gravid female.

***Xiphophorus xiphidium* male, Rio Purificacion.**

***Xiphophorus couchianus* male from Apodaca, Mexico.**

ALL PHOTOS BY LOTHAR WISCHNATH.

***Xiphophorus variatus* female.**

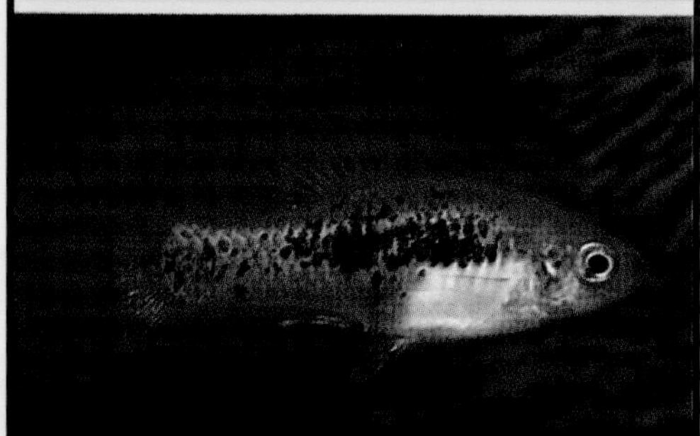

***Xiphophorus variatus* male from the Rio Nautla.**

***Xiphophorus variatus* female from the Rio Axtla.**

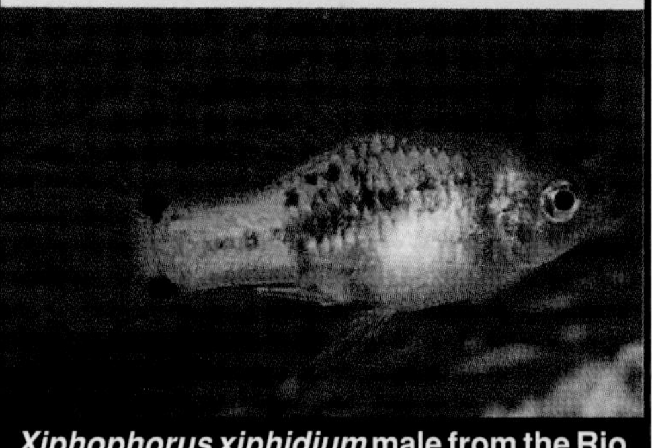

***Xiphophorus xiphidium* male from the Rio Purificacion.**

ALL PHOTOS BY LOTHAR WISCHNATH.

***Xiphophorus variatus* male from Mexico.**

***Xiphophorus variatus* male from Mexico.**

***Xiphophorus variatus* male from the Rio Axtla.**

Xiphophorus xiphidium male, from the Rio Santa Engracia.

***Xiphophorus xiphidium* (Hubbs & Gordon,1932). Swordtail Platy or Purple Spike Platy**

The male is up to 4 cm (1.6 in) with the female being much larger, to 6 cm (2.4 in). Color is a variable blue-green above becoming paler below. The male may sport blotches of black above and behind its gills. The female may have dark stripes on her anterior flanks. Both sexes have a dark gravid spot, and one or two dark patches on the peduncle. There is a longitudinal line of a variable orange color at the level of the lateral line.

The male has small extensions to its lower caudal fin, thus the common name of swordtail platy. This species can tolerate a temperature a little higher than most others in the genus, but will do well at the preferred settings for platies, about 72°F-73 (22°C). It is perhaps the most delicate of the platies from the hobbyist's viewpoint.

***Xiphophorus xiphidium* male.**

***Xiphophorus couchianus* female**

DOMESTIC VARIETIES

When compared to the immense array of mutational forms, the wild platies are positively somber and plain. There are now so many variants that beginners, indeed even experienced enthusiasts, can be overwhelmed by the number of names applied to them.

It would be impossible to describe all the domestic forms because the potential permutations are infinite. Some of these can be bred consistently, others cannot. They are the result of the random uniting of genes to create individual fish that may range from ugly to breathtakingly beautiful. The end result, while being a nightmare for the specialist breeder, means that you as a new hobbyist can select from a wonderful range of forms and colors.

THE BASIC GROUPS

Platy variants and their hybrids can be classified using three simple divisions. There are the color varieties, the patterned varieties, and those that show fin modifications—the dorsal and caudal fins being the ones most visually affected. Each of these variations are inherited independently. This means that combinations of each mutational group can appear in a single fish. Certain varieties are not fertile because if the mutation results in a drastic modification of the gonopodium they are unable to transfer their sperm to the female. In other instances it is possible that all the young from a mating may be males. The reason for this will be explained in the breeding chapter that follows.

COLOR VARIETIES

The base colors seen in these fish can be yellow, gold, red, green, brown, blue, and black. Additionally, there is the albino, which is a fish devoid of color pigment, and with pinkish to red eyes. The albino may be pure in the genetic sense of displaying no color whatsoever, or it may display some pigmentation in certain parts of its body where these are not the result of the dark melanins.

Each of the colors is possible in a vast range of shades, which can vary within the same fish. It is in the reds and golds that trade names are most commonly applied. In reds there is coral, brick, blood, and velvet. Gold may be marigold, sunset, sunburst, or simply golden. As with any color shades, what one person regards as one color someone else may perceive as being another, so shade names are rather subjective.

Most of the color varieties of platies were developed by Dr. Myron Gordon and his student, Dr. Herbert R. Axelrod at New York University's Genetics Lab atop the N.Y. Museum of Natural History.

ALL PHOTOS BY EDWARD TAYLOR.

Blue wagtail female.

Blue female platy.

Golden wagtail platy.

Black platy.

Green platy female.

Modified green platy male.

Blue variatus male.

Red wag female.

PHOTO BY EDWARD TAYLOR

Red tuxedo female.

COLOR PATTERNS

Given the fact that there is a potential for more patterns than colors, you may see all manner of names applied to them. However, the following can be regarded as basic, and are standard in the hobby.

Tuxedo: As the name suggests, the tuxedo is a bicolored fish. The posterior part of the body is black, while the anterior area is any other color. The quality of tuxedo is extremely variable, ranging from impressive to debatable in some instances.

Wagtail: In this pattern the rays of the dorsal and caudal fins are black.

Salt & Pepper: The body is liberally covered with small light and dark spots as might be expected with this name. It is a variation on the variegated pattern.

Variegated: The body is covered with dark blotches of variable size in no particular arrangement. In nice examples the pattern is impressive, in poor ones it will be the opposite. If a dark spot is centered around the peduncle, and the tail carries round blotches on its upper and lower edges that connect with the

Two male golden wagtail with red saddles.

PHOTO BY EDWARD TAYLOR.

PHOTOS BY EDWARD TAYLOR.

Red salt-and-pepper platy male.

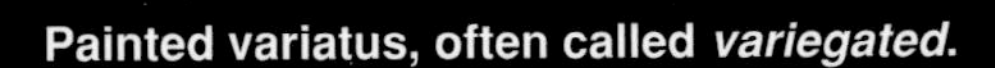

Painted variatus, often called *variegated.*

peduncle spot, the pattern formed is called the "Mickey Mouse" pattern.

Rainbow: This needs little description. It is a fish displaying a whole range of colors. Very much a chance pattern.

Berlin, Hamburg, and **Weisbaden:** Named for German towns, each with a variation on the basic black color. The scales are black with green edges, or black predominates with a color (red, orange, yellow, or green) within the scales, black spots mark a red body, or the upper part of the body has a different color.

PHOTO BY EDWARD TAYLOR.

A male rainbow platy produced by Imperial Tropical Fish Farm.

Pineapple: In this variety the scales have the appearance of the outer leaves of the pineapple.

Bleeding Heart: In this pattern the red is centered around the flanks and underparts in the area

Painted wagtail platy produced by Bill Liles Tropical Fish .

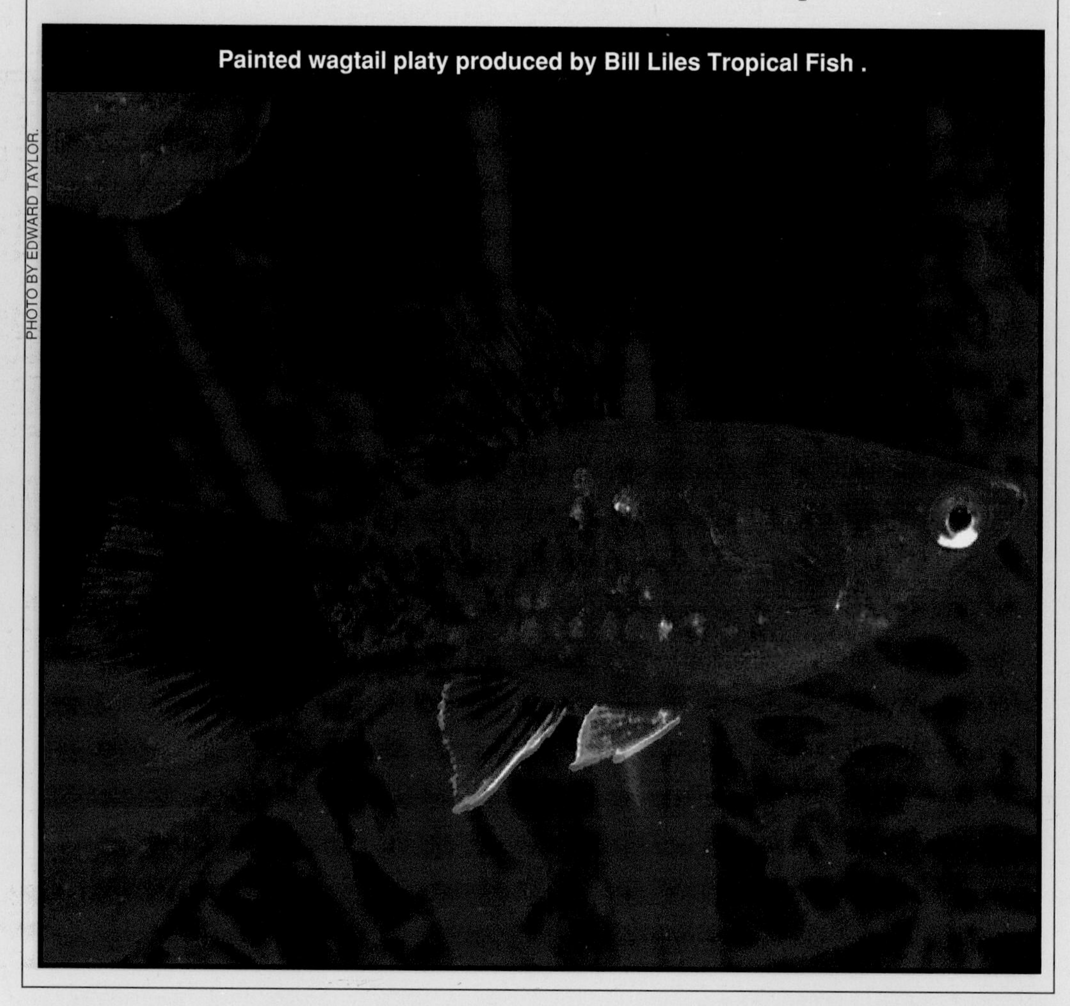

PHOTO BY EDWARD TAYLOR.

PHOTO BY DR. HERBERT R. AXELROD.

The first bleeding heart platies were exhibited at the New York Zoo and were developed by Drs. Myron Gordon and Herbert Axelrod.

Pineapple wagtail platy produced by Imperial Tropical Fish Farm.

PHOTO BY EDWARD TAYLOR.

of the heart, fading to orange as it moves onto the rest of the body.

With the exception of the wagtail, the patterns seen in domestic variants of the platy are only partly under breeder control. By this is meant that the pairing of two excellently patterned individuals by no means ensures that the progeny will be as well marked as their parents. The reverse is equally true, in that superb examples can be produced from very mediocre patterned parents.

FIN TYPES

Although platies do not as yet exhibit as many variations to their fin shapes as can be seen in the guppy, those that are well established are impressive in good examples. These are as follows:

Hifin: In this variety, also known as the Simpson Hifin, the dorsal fin is elongated to a variable degree. In some examples it can be so long that the fin rays are incapable of maintaining it in an upright position—part of it trails over the sides of the body.

PHOTO BY DR. JOANNE NORTON.

The genetic base to the patterns is complex. But, even with this reality, from a practical viewpoint it is still best to pair well marked examples and develop strains on the basis of selecting the best of the type desired from the offspring. By such policies some breeders have been able to move toward a situation where the number of well marked individuals has risen over a given term.

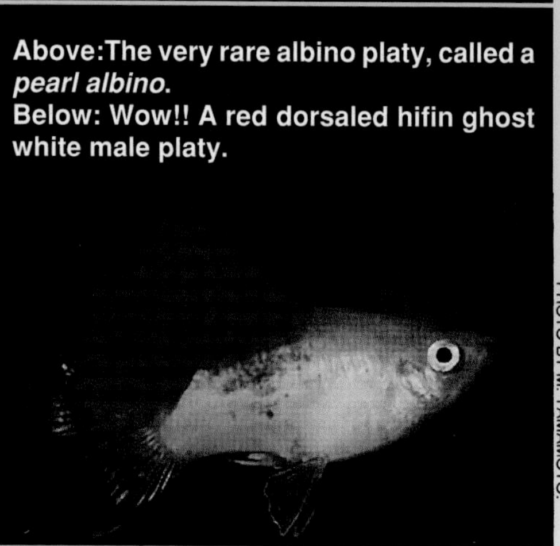

Above:The very rare albino platy, called a *pearl albino*.
Below: Wow!! A red dorsaled hifin ghost white male platy.

PHOTO BY M. YAMAMOTO.

Red wag pintail female platy in good color.

Pintail: In this variety the center rays of the caudal fin are extended to create what appears like a second sword half way up the tail. This extension may be longer than the bottom sword in short sword varieties.

A final comment is worthy of mention with regard to domestic forms in general, and especially when they involve enlargement of the fins. When any mutational form is developed it will tend to be somewhat weaker in constitution when compared to its wild ancestors. Such fish will be rather less tolerant to conditions that are less than they should be. The greater the movement away from the wild state, the more delicate the fish.

With specific reference to fin modifications, these will be much more fragile and prone to bacterial or fungal attack if the water conditions deteriorate. They are also more prone to being nipped by other fishes. Given these facts it is very important that if you keep the more exotic forms of platies you must pay extra special attention to the water conditions.

Red hifin male platy. Fish bred by Victor Meyer.

PHOTO BY EDWARD TAYLOR.

BREEDING

The poeciliids are both very easy and extremely difficult fish to breed. This apparent contradiction arises based on how you define the term breeding. If this is regarded as the ease with which platies will procreate in an aquarium they are very easy to breed. If a male and a female are present they will need no help from you to produce offspring.

However, if breeding is defined as the creation of individuals that resemble their parents in color, pattern, and fin form they are very difficult to breed. Instructions on the selective breeding of this group of fishes could not be discussed in a single all embracing chapter, so lets look at the subject from a purely practical viewpoint. This said, a few initial comments on selective breeding will be made so that you can appreciate just what problems are involved in attempts to produce given colors and patterns.

If you find these challenging then you are referred to more specialized works that detail the complications of genetics in fishes. A good start in this latter area would be to obtain *Genetics for Aquarists* by Dr. J. Schroder, published by TFH Publications.

PROBLEMS WHEN BREEDING LIVEBEARERS

Many of the livebearers, which include the platies, have developed a survival strategy known as superfetation. The female is able to store sperm packets from a single mating by a male. She is then able to use these to fertilize successive broods (up to nine) for a number of months. This is why beginners with only a single female of one of these species are surprised to find that she may suddenly produce offspring—the mating clearly took place before they acquired the fish.

The problem this capacity creates is that if females are not kept apart from males from an early age, they will be mated—probably by an unwanted color variety, or even species. Most poeciliids will hybridize, so if you have guppies or mollies in a community tank with platies, it becomes pot luck who mates with who. If you are trying to develop a strain of a given color or type you can immediately see the problems resulting from even a single unwanted mating.

A magnificent red comet platy female.

PHOTO BY DR. HARRY GRIER.

SEXING PLATIES

If you can be satisfied initially with just producing more platies you should keep the females away from any other male fish of this family. You must also be sure that the females could not have mated before you obtained them. If your dealer cannot assure you on this point it would be prudent to obtain your first females from an established breeder.

Platies are sexed on the basis of the males. A male swordtail develops the sword extension to its caudal fin, as well as a gonopodium, which is a modification of the anal fin via which sperm packets are deposited into the female. The male platy has no sword, so it's the gonopodium you will be looking for.

This is variable in the age at which it has developed enough to be recognized. In swordtails, maturity in the males may take rather longer than in platies—up to nine months in some individuals. However, as a general guide you can usually sex them when they are two months old — or even younger.

MATING METHODS

If you have a mixed community of only swordtails or platies, or both, mating will take place naturally on a random basis. If you wish to be selective you must keep young fish separate from the outset. At this time you may still not be able to sex the young. If you have a mature male available he can be placed with the youngsters. You will have to work on the premise that developing males within the youngsters are not normally able to fertilize the females until a short period after their sex is apparent, although this is not always the case.

The mature male will mate with the females, so you have a very good chance that he will be the father of most, if not all, of the young produced. If you have young females of the desired type and color, these can be kept with one or more males of similar, or the desired, type. After each brood is born you will remove these (or the breeding adults) and maintain the same parental stock together.

Eventually, any sperm packets the female had before she was placed with the desired male(s) will be used up. You will recognize this fact in the appearance of the young of the type that resembles the father, assuming he was of a different color and pattern. As you can see, the problem is not in breeding these species, but in being able to control which males mate with which females. In order to be a selective breeder of these, and most other livebearers, you will need to have an extensive set up that includes quite a lot of breeding and rearing tanks so that you are able to keep the different types separated.

INCUBATION TIME & BROOD SIZE

The time lapse between the female's eggs being fertilized and the young being born is about 29-33 days, depending on the water temperature (the warmer it is the shorter the incubation period). Brood size is very variable because it is influenced by numerous factors. These include the pH, temperature, fitness of the parental stock, their age, and their genetic background. Generally, the first litter will be small, then successive broods will get larger, before falling numbers indicate that the female is getting old. As a guide, the range is 20-200, with 40-60 being typical.

CARE OF THE GRAVID FEMALE

Before, and throughout, her period of "pregnancy" the female should be given increased rations of high protein foods. It is best if some live foods can be given. Most certainly you should not feed just flakes as these are not conducive to good brood size, or to top conditioning. The male, too, should be in super condition before he is bred, otherwise the vigor of his sperm will be affected.

As the incubation period progresses you will see that the female develops a dark area on her abdomen; this is called the gravid spot. Shortly before the young are born you may even be able to see them through the stretched abdominal skin of the female. Once the female is clearly pregnant there is no need to keep the male with her.

PROTECTING THE FRY

Having successfully mated the female your thoughts must now turn toward protecting the young once they are born. If this is not done the parents will have no qualms about having them for dinner as they are released from the female! On this aspect you have one of two options.

PHOTO BY BURKHARD KAHL.

Extremely well colored and uniform red wagtail platies.

Breeding Traps: These are commercially made and sold by your local aquarium dealer. They are basically plastic cages in which the female is placed just prior to her giving birth. The tiny young are able to swim out of it via slits in the trap which are too narrow for the female to go through.

Nursery Tank: In this small aquarium, plenty of vegetation is included so that as the young are born they are able to quickly retreat from their potentially dangerous mother. Some young will no doubt be eaten before the mother is removed, but most will survive.

The advantage of the nursery unit is that the female is less likely to become stressed—a condition that could result in a reduced litter, or no litter, depending on the temperament of the female. A compromise between these two alternatives would be to bond two Plexiglas (Perspex) divider panels in a "V" configuration about one-third down the length of an aquarium. Leave a gap of about 3-4 mm at the point of the "V" for the young to swim through. The breeder end can be planted and simply filtered so that the female is content. As the young are born they can swim, via the gap in the "V", into the safe zone. Here, they can easily be gathered and transferred to a well planted nursery tank.

RAISING THE YOUNG

As the young are born they may initially sink in the water but will soon swim toward a source of light, which is why a well illuminated tank should be used for rearing. It should contain a good supply of floating vegetation under which they can feel secure. They will begin feeding very quickly. A supply of fry food (micro-size brine shrimp and similar live foods, or liquid or powdered food) should be on hand.

They will need a number of feedings per day, and an adequate amount to ensure that all the young get their share—not just the bolder, more aggressive individuals. The substrate of a nursery tank should be of a simple type so that there is no risk of water pollution. By keeping things simple you can siphon the substrate daily. A foam filter will suffice for your young, but be sure to keep your eye on the water quality. As the young grow quickly there will be a need to increase the oxygen content—or to transfer the young to a larger tank.

As with any breeding program, there are going to be some sickly or malformed offspring. A number of these will die soon after birth, but those that survive should be placed into a display tank which contains adults who will quickly devour them. If this does not appeal to you they should be removed and placed into the deep

DRAWN BY JOHN QUINN.

Floating baby trap with a female platy on each side.

freeze part of your refrigerator, this being the most humane way to destroy a fish.

You should grade the healthy young continually over the first months of their lives, selling all but the very best, which you will want to retain for subsequent breeding.

BREEDING RECORDS

If you take breeding seriously you must keep accurate records as these will be invaluable in your selection program. Never forget that it is very poor policy to retain only the most beautiful fish while disregarding those less appealing individuals that have greater vigor and a proven ability to produce good brood sizes of youngsters that have an excellent health record. If this aspect is not made your prime objective, you will be removing the compensatory replacement for natural selection. Successive broods may start to become weaker, more prone to ill health, and display poor breeding vigor.

KEEPING YOUR FISH HEALTHY

Health problems and diseases in fishes can be divided into two broad categories—external and internal problems. Your chances of remedial success with the former are infinitely better than with the latter. This is because external problems can more readily be seen before they reach a dangerously advanced state. Internal diseases, apart from being much more complex, are usually in an advanced non-treatable state by the time they are evident.

Even then, the problem comes in the diagnosis. Often, this can only be effected by the use of microscopy of blood samples, or an autopsy of the affected organs in the dead fish. Most aquarists will not pay the cost for these veterinary services, and even if they do there is not always a guarantee that the problem can be diagnosed. If it can, there may be no cure anyway, so the prognosis for treating internal problems is hardly very good.

There is another set of problems when attempting to treat fishes. Unlike the situation in birds, mammals, and reptiles, the treatments can of themselves have a very negative effect on the environment in which the fishes live. Potential benefits gained from the medicine may be negated by the damage they do to the water quality and all the organisms and plants in it.

For example, while a medicine may cure a fish, the quantity needed may be more than enough to wipe out the entire colony of aerobic bacteria needed to convert nitrites to nitrates. It may also be enough to wilt or kill the vegetation. In the former instance, should this happen, the fish could die anyway, so the whole treatment would have been wasted.

If your filter system features charcoal and similar chemical filters these may adsorb medicines, so they are less than effective. Each of these problems presents the aquarist with complications well beyond the actual problem or disease they are trying to overcome. Where fishes are concerned, the adage that "prevention is better than cure" really does make eminent sense!

Within this short chapter a discussion of specific diseases is not possible, so we will look at preventative husbandry. This has more practical advantage to you than attempting to be an expert on a legion of problems and treatments it is hoped you will never experience. With this sound base to work from you can, if required, purchase books that deal more specifically with known diseases and their treatments—most of which will revolve around what you learn here.

WHY PROBLEMS AND DISEASES CAN BECOME RAMPANT

Most problems in aquaria can be traced to a number of identifiable causes that, largely, can be overcome with sound management. Those that are not, and result in a problem, can be minimized by the correct response, both in timing and action. There is thus lots of good news as well as the bad already discussed! The causes of health problems stem from one (or more) of the following:

1. **Poor water quality:** If the pH, hardness, nitrite, phosphate, oxygen, carbon dioxide, chlorine, and other element or compound factors are not as they should be, the fishes will have problems. Under this heading can also be included dirty water, and water of an inappropriate temperature. All of these factors intermingle with each other to result in differing degrees of polluted water.

2. **Overcrowding:** It is probably true to say that over 80% of all display aquaria are overcrowded when compared to the volume per fish situation found in wild habitats. This means that even if the aquarium water is of excellent quality, there is still a high degree of risk created by the close proximity of the fishes to each other in a body of water that is not being totally replaced every few seconds as it would be in a natural river or stream habitat.

3. **Late recognition** that a problem already exists: If an aquarist is not observing his or her stock daily, there is the obvious potential for a problem to go unnoticed until it is so obvious it cannot be missed.

4. **Slow or no reaction to a problem:** Very often owners will wait for a period of time to see if what appears to be a minor problem will clear up by itself. This is a dangerous assumption in a limited volume of water, which is what even a large aquarium actually is. If you work on the premise that if a problem is seen it should be dealt with immediately, you will minimize its opportunity to get worse and spread to the rest of the fishes.

5. **Incorrect diagnosis:** It is quite ineffective, and may even be counter productive, to attempt to treat a problem unless it has been correctly diagnosed. Most treatments will stress your fishes to a greater or lesser degree. They certainly do not need this burden unless it is likely to alleviate the condition. Never guess at problems, but try to diagnose them correctly based on observable facts. This done, seek the advice of your aquarium dealer, an experienced aquarist, or your vet.

DIAGNOSING A PROBLEM

If a fish is unwell this will be displayed (if at all) via one, or both, of two ways. There will be visual (clinical) evidence, or there will be behavioral indicators. Based on these, a course of action must then be determined and put into operation as soon as possible. Diagnoses should take into account both what is observable, and what is not. By the latter is meant that you should consider all factors that may have had an influence on the situation you are now observing. This is known as anamnesis, the recollection of all events that precede an illness. We will look at each of these three subject areas.

Clinical Signs: I am sure that you are well aware what a healthy fish should look like, so it may be assumed that any other appearance is abnormal, thus a problem. Specifically, the following are clinical signs of illness: Cloudy look to the eyes, any form of swelling on the head, body, or fins, sunken eyes, pop eyes, cuts, abrasions, or ulcerations, cotton-wool or fluffy like growths attached to the fish, fins that are ragged and appear to be rotting, blood streaked fins, slimy secretion of mucus covering the scales, white or yellow spots on the body and fins, scales standing out from the body rather than laying flat, any form of organism attached to any part of the fish (this is where a good hand lens is useful), redness of the gills, very poor coloration, and emaciation.

Behavioral Signs: Rubbing against rocks or other objects, swimming at an abnormal angle, inability to maintain a given position in the water (rising or sinking), gasping at the surface, lying on rocks or the substrate, fins outstretched or clamped close to body, frequent sudden darting through the water (but all fishes will dart if startled), obvious lethargy, resting excessively in hidden positions behind rocks, in caves, or near plants, and disinterest in favored food items.

Anamnesis: Were any other fish ill before the present one? Were the signs similar or different? Is the present patient(s) a recent arrival? Was it quarantined? Have other fishes from the same source presented problems? Has the water quality or temperature changed lately? Is there a possibility toxins might have found their way accidentally into the water? Have new rocks, gravel, or plants been introduced recently, and what was their source? What live foods are in the diet, and what is their source? When was the last time you tested for all(!) the major parameters of water quality?

With all the detective work complete you are now more able to try and diagnose the problem and pinpoint its potential source. Your first course of action is to make complete readings of the water quality and properties. This of itself may indicate the problem and remedy. If only one or two fish are affected it would be wise to remove them as quickly as possible and treat them in a hospital tank.

If many fishes have a problem you may have no option but to treat them in the aquarium, taking the risks that have already been mentioned earlier in this chapter. This being so, you should remove charcoal and similar filters. Increase aeration via air stones rather than via any increase in the flow of biological filter systems. This would accelerate the effect of medicines on the beneficial bacteria by increasing the water flow through the filters.

TREATMENTS

Some medicines have now been developed that are selective in what they kill. They may not have a disastrous effect on beneficial bacteria or plants. Likewise, others have a broad spectrum of efficacy, so that they will destroy a range of parasites rather than be selective. This has the benefit that even if you have diagnosed the wrong parasites the treatment may well be effective.

Some treatments can be used concurrently with others, but some cannot—they might double up on a common chemical, or interact with it to be dangerous, or non effective. Another consideration is whether to remove some (or all) of the rocks and plants from the aquarium. Plants can be replaced by plastic ones so that the fishes still have somewhere to hide. They may otherwise become stressed, which would prolong the healing time. In effect, you must try and convert the display tank into a hospital tank when one of these is not available.

Hobbyists can take advantage of many remedies and preventatives which have been formulated especially for use with platies and other tropical fishes. A great number of the most common fish diseases are easily cured with inexpensive, non-prescription preparations available at your local pet shop. Photo courtesy of Aquarium Products.

Many different remedies, preventatives and tonics are available at pet shops. Photo courtesy of Jungle Laboratories

Once the treatment appears to be making headway your next decision, depending on the severity of the problem, will be whether or not to set up a holding tank and strip the display unit down totally and start again from scratch. The facts that might dictate this are that many diseases can survive as spores in the water, so they are not killed by the treatments. Thorough disinfecting of all equipment coupled with the replacement of some decorations may be preferable to the risk of a fresh outbreak of the disease.

To complete this unhappy scenario you may also need to ponder whether the recovered fish and others will remain as carriers. Hopefully, you will never be faced with these drastic decisions, but they are a reality of life so it is well that you fully understand all implications of a disease. By so doing, you may be less inclined to risk overstocking your aquarium, and more amenable to purchasing a small hospital tank so that you can take prompt action the minute a problem is observed.

QUARANTINING FISHES AND PLANTS

From the foregoing discussion you will appreciate just how important it is to obtain only very healthy fishes, and then never to overstock the aquarium. A reliable source of fishes is imperative, but even the best of these can sell infected stock unbeknown to them. Your backup insurance is to quarantine all new arrivals. This becomes progressively more important the larger and more well established your display tank is.

The quarantine or hospital tank need only be a small tank. It should be sparsely furnished. Marbles can be used as a substrate, and a few plastic plants, as well as a small flower pot to provide a cave-like structure for the fish to retreat to, can be included. A simple foam filter should be sufficient, plus a small air stone. By using synthetic furnishings you minimize the risk of pathogens being introduced, and these items can all be thoroughly disinfected and rinsed after each use.

The quarantine period should be 14-21 days, which should be enough for any incubating problems to manifest themselves. During this period you can routinely treat for parasites, as well as adjust the diet to the one you use. You can also acclimatize the new arrivals to the main display water by introducing some of this during the later stages of quarantine. All these measures will make the change from one tank to the other less stressful for the fishes, and safer for the main ecosystem.

When reading any health care chapter it can seem as though there are enormous difficulties confronting the aquarist. But this is only ever the case if routine matters of maintenance are relaxed, and common sense ignored. Check your fishes daily and make regular trips to your chosen aquarium store. Its staff may prove to be real life-savers should a problem of consequence become a reality. You do not get this sort of service advice from supermarkets that want your business, but not your problems!

SUGGESTED READING

H-1077

H-1090

TS-131

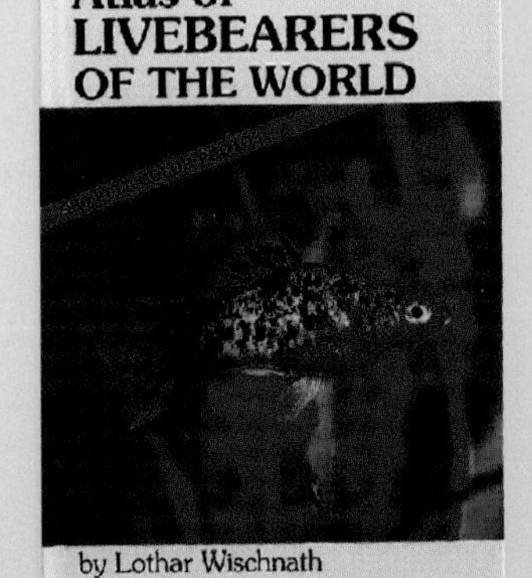

TS-180

H-1028